RAILWAY PROPERTY.

A TREATISE

ON THE

CONSTRUCTION AND MANAGEMENT

OF

RAILWAYS:

DESIGNED TO AFFORD USEFUL KNOWLEDGE, IN A POPULAR STYLE, TO THE HOLDERS OF THIS CLASS OF PROPERTY; AS WELL AS TO RAILWAY MANAGERS, OFFICERS, AND AGENTS.

BY JOHN B. JERVIS,

CIVIL ENGINEER.

Facility of Communication, in Social, Commercial and Political Intercourse, is a distinguishing Index of Civilization.

PHILADELPHIA:

HENRY CAREY BAIRD,

INDUSTRIAL PUBLISHER,

No. 406 WALNUT STREET.

1872.

PREFACE.

THE following pages have been written with a view of giving to railway proprietors, in a popular form, the character of their property. In this I have aimed to give some impression of the public importance of railways, but have mainly discussed the methods of business involved in their construction and operating management. In regard to construction, it has been my design to give so much of particulars as will enable an intelligent business man to form a judgment of the propriety of the several proceedings, and enable him to decide on their wisdom and expediency, and at the same time afford the junior engineer hints and outlines that may be useful for his consideration. At this time, the operating management, providing as it does for the care of near thirty thousand miles of railway, is far more important than that for construction, in which there is comparatively little doing.

The care of more than one thousand millions of dollars invested in this kind of property, all must admit, is of very great importance to the proprietors. In addition to tracing out, and presenting the order of business, it has seemed to me necessary to point out the danger to which the property is peculiarly exposed, from infidelity to trust. Though I regard the ordinary and current management of railways as very far from perfect—far from what it will be by and by, still there can be no doubt that the failure of railway property to remunerate the proprietors, has, to a large extent, resulted from unfaithful management, and therefore I have endeavored to show the working of this evil, and, so far as practicable, the remedy.

If, in the following discussion, the prospects of this class of property do not look as encouraging as could be wished, it must not be attributed to any desire on my part to disparage it; but to the evils that environ it, and the difficulty of wholly removing them. It will be seen that I do not regard this kind of investment as well suited, in general, to small proprietors, so situated that they can exercise no control, and who are exposed to the danger of having their property managed by unfaithful men, who seek to make the institution subservient to their own interest, rather than to that of the proprietors.

All that is solicited from the reader of the following pages, is a candid and impartial survey of the subject. Doubtless I have advanced some opinions that will not harmonize with much of the existing management on many railways; but it is confidently believed that time and experience will fully demonstrate the soundness of these opinions, and ultimately introduce a practice that will greatly economize transport, and extend the usefulness of this eminently beneficent improvement in the means of intercommunication.

The proficient in railway management will find much that may appear to him commonplace; but he will readily see, that I have written to give useful information, chiefly to those who have had less experience than myself, and who, I hope, will derive some benefit from the labor I have bestowed on a subject of general, as well as of individual interest.

CONTENTS.

RAILWAY PROPERTY.

INTRODUCTION.

The vast amount of funds (about one thousand million dollars) invested in American railways, claims the especial attention of those who have furnished the means for their construction. Vast, indeed, if we consider the infancy of our country, and the large demands for capital, indispensable for other purposes, namely, to clear the forests, to drain and fence the lands, to erect dwellings for the people; barns, stables, and sheds for the storage of farm products and shelter for animals; shops, manufactories, tools, implements, and machinery for artisans and mechanics; and for wharves, warehouses, stores, shops, shipping and other appurtenances of commerce; all which must be provided in a new country from the surplus beyond what is necessary to supply current wants.

It is now little over a quarter of a century since railways were introduced as a means of promoting the general intercourse of men. The opening of the Liverpool and Manchester Railway, completed in 1830, may be regarded as substantially inaugurating this improvement. Previous to this, something had been done, as on the Stockton and Darlington Railway; but the wants of the general traffic in passengers and freight by locomotive power, were not fully met until the time above stated. The success that attended the grand experiment of George Stephenson, Esq., Civil Engineer, was not lost on the mind of the civilized world, and railway improvement, especially in England and our own country, was pushed forward with extraordinary zeal and energy. So far as a beneficial result was anticipated, in advancing the commercial and social intercourse of men, the strongest hopes have been realized.

It is not regarded as important at this time, after so much has been demonstrated by practical operation, to say much in relation to the usefulness of railways. It may be well, however, to call attention to the opinions formed at a period when the experience of men could see, in a near view, the advantages they afforded in contrast with the means of communication that existed before they were known. The following quotation, from the "Westminster Review," for December, 1845, article 7, presents in vivid language the view entertained on this subject in England at that time. "Let us clearly understand our position. We have arrived at a new epoch in the history of the world. A new element of civilization has been developed. As was the invention of letters, as was the printing-press, so is the railway in the affairs of mankind. It is a revolution among nations. A moral revolution as affecting the diffusion of knowledge, the interchange of social relations, the perpetuation of peace, the extension of commerce; *and a revolution in all the relations of property.*

"We refer, by the latter observation, to the cheapening of all kinds of commodities by the facilities of carriage and the saving of time on the part of producers, afforded by railways, and especially to the influence of railway upon the value of houses and lands . . . hence the demand for railways to connect every town of the United Kingdom . . . is a real bonâ fide want of society, which sooner or later must be supplied." This view, presented in 1845, has lost none of its force by subsequent experience. No reflecting mind, intelligent enough to contrast the condition of the world before and since the introduction of railways, can fail to see the force of the language held by the "Review." If attention be given to particulars, it will only show its general conclusions to have been founded on the most practical developments of the improvement.

"The diminished cost of transport invariably augments the amount of commerce transacted, and in a much larger ratio than the reduction of cost; so that, in fact, although a less amount of labor is employed in the transport of a given amount of commodities than before, a much larger quantity of labor is necessary, by reason of the vast increase of commodities transmitted. The history of the arts supplies innumerable examples of this. . . . The moment the first great line of railway was brought into operation between Liverpool and Manchester, the traffic between those places was quadrupled. . . . Improvements in transport which augment the speed, without injuriously increasing the expense or diminishing the safety, are attended with effects similar to those which follow cheapness. . . . Numerous classes of articles of production become deteriorated by time, . . . admit of transport only when they can reach the consumer . . . in a sound state, . . . as various articles of food; . . . animals of every species driven to market on common roads, were proved to suffer so much from the fatigue of the journey, that when they arrived at market their flesh was not in a wholesome state; . . . sheep frequently had their feet literally worn off, and were obliged to be sold on the road for what they would fetch. Extensive graziers declared that, in such cases, they would be gainers by a safe and expeditious transport for the animals, even though it cost double the price paid by the drovers." (Lardner, on the Economy of Railways, pages 7, 8 and 9. London: 1850.)

In little over a quarter of a century—1826 to 1858—there were constructed and put in operation in Great Britain about 9,000 miles of railway, at a cost of say fifteen hundred million dollars. A high confirmation of the opinions and statements above quoted.

"At the present time nearly 9,000 miles of railways have been completed in the British Isles. . . . It appears from the published returns of traffic upon railways for the year 1856, that 129,315,196 persons travelled 1,822,049,476 miles, and paid nearly £11,000,000 in fares; 10,450,625 cattle, sheep and pigs, . . . 23,823,930 tons of merchandise, . . . "and 40,938,575 tons of minerals were carried. The total receipts on the railways for the year amounted to £23,165,493." (See "Edinburgh Review" for April, 1858, Article 4.)

Compared with the previous history of works and improvements in the means to facilitate the commercial, political and social intercourse of mankind, the railway truly marks an "epoch." The superficial area of country on which the larger expenditure above quoted has been made, is 121,000 square miles; but little larger than the area of the States of New York and Virginia; and three-fourths of these roads are in England on an area of about 60,000 square miles, and afford nearly a mile of railway to ten square miles of territory. Most of the British railways have a double track; single track railways occur only as rare exceptions.

The foregoing is a sketch of what has been done, and of the views entertained on the subject of railways in Great Britain. And here it must be observed, the country is surrounded by coast navigation that has always been extensively employed for the transport of persons and property—the interior provided with near 3,000 miles of canal navigation, and 27,000 miles of McAdamized turnpike roads, of the best quality.*

If railways have produced so great results in a country like England, largely provided with the best means of transport known previous to their

* See "Westminster Review," December, 1845, Article 7.

introduction, how vast must be their beneficial influence on a country like our own! The United States has many advantages of coast, lake and river navigation, and a considerable extent of canal navigation. In many of these, there are frequent interruptions from drought, and all in the higher latitudes are, during a portion of the year, locked up by frost. At best, they leave broad districts of country between them, that have no other means of transport than earth roads, depending mainly for their quality on the character of the natural soil on which they are located; which are at all times inferior, and often unfit for the transport of even moderate loads. No such thing as rapid travelling could be expected, and the stage-coach passenger must be content to jog on at the rate of two to five miles per hour, according to the dry or wet season. This was peculiarly the condition of transport over the deep loamy soils of the western States. At this day the memory of many men will call to mind the hardships of a winter journey between New York and Albany, previous to the introduction of railways. I have a vivid recollection of several such journeys, and of one in particular, that occupied two successive nights, and parts of three days' continuous journeying. Now the jaunt is made in five hours, with nothing worthy the name of fatigue, and at comparatively trifling expense. A journey from the seaboard across the mountains to the western States, was only encountered by the vigorous and hardy, and by them only on high inducements, that would justify the time, expense and fatigue involved in such an undertaking. To the invalid, there was little prospect of even moderate journeyings. As an illustration of this, I refer to an instance in my own experience. In March, 1836, it became important for me to make a journey from Albany to Rome, in the State of New York, with an invalid wife, who was too feeble tc bear stage travelling. At that time, the railway was opened from Albany to Schenectady (16 miles), and so far I had its benefit. It required four days to reach Utica (96 miles), at an expense of over $50 for four persons. At this day, such a journey could be performed in four hours, at an expense of $8. On another occasion, an invalid friend, too feeble to walk a dozen paces, was put into the railway coach at Rome, New York, and in nine hours made the journey to New York city, 250 miles, without suffering inconvenient fatigue. Doubtless, many instances might be added, to show the beneficent influence of railways, in promoting the comfort and convenience of the aged and infirm.

By the old system of stage-coach travelling, the average distance travelled in 24 hours cannot be placed higher than 100 miles; the expense of fare, $5, other expenses, $1 50, total $6 50. A journey of 500 miles would require five days, and an expense of $32 50. By railway, the journey would require one day, or, 24 hours, at an expense not often exceeding $15 (in some cases not exceeding $11), thus saving $17 50 in expense and four days in time on a journey of 500 miles. On some routes the contrast is greater, both in time and expense. Within two years, I was travelling on the railway between Buffalo and Cleveland, and while the train stopped at a station, I heard two young gentlemen in discussion, complain sharply at the slowness of our progress; I inquired what speed they supposed we were making, to which they replied they did not know; I said, we appeared to be on time, and that would give us about eight hours from Buffalo to Cleveland, and contrasting this with experience I had had on the same route before the railway was in operation, and at about the same season of the year (winter), when it had occupied four days of very dilligent travelling to accomplish the same journey, our progress on the railway appeared very satisfactory. The comparative time of the railway was about one-tenth. This incident is mentioned for the benefit of those whose experience does not enable them to appreciate, by comparison, the change the railway has wrought in the means of easy and expeditious transport.

American industry, in agriculture, commerce, and manufactures, has received an impulse from railways, that has carried them beyond the most sanguine anticipations, greatly advancing "the value of houses and lands." The saving in time, expense, and fatigue of travelling has increased personal intercourse, it is believed, more than tenfold. This is manifest in the vast number of short journeys; and those of 100 to 1,000 miles are now undertaken with comparatively little thought, not certainly more than would have been given under the old system of travelling for journeys one-fifth as long. Consequently, our political, commercial, and social intercourse has been greatly extended. In the interior districts, and especially in the western States, their influence on agriculture has been most obvious. Farm products have advanced in value from 20 to over 100 per cent., and this range will hold good for a large portion of our territory.

By opening and enlarging the facilities of intercourse—giving increased value to the products of labor—introducing articles of necessity and comfort, and the means of intellectual and moral culture, it has made much of our country, that was before secluded, pleasant and desirable for places of abode; thereby greatly promoting civilization.

No great effort of mind is necessary to see in the railway, a vast labor-saving machine, diffusing its benefits to all classes and conditions of men; increasing the value of each of their own products, and cheapening the cost of articles of necessity and comfort obtained from distant places.

Those who desire to examine particulars of the history of intercommunication, and the contrast between the facilities that existed prior and subsequent to the introduction of railways, are referred to "Lardner's Economy of Railways," London, 1850, where they may find much that is interesting on this subject.

By the last report of the postmaster-general, it appears that the United States mail was transported in June last (1858), over 24,431 miles of railway. Doubtless, there was some railway opened at that date on which the mail was not carried, and some opened subsequently before the close of that year, making an aggregate of 26,000 miles of railway in this country in operation at the close of 1858. The total cost may be set down approximately at *one thousand millions of dollars*. Hunt's "Merchants' Magazine," for September, 1858, states the gross receipts for 1857 at $98,949,600, and for 1858 (which must have been partially estimated), at $106,013,600, and the net income for 1858 at $44,589,400; equal say 4½ per cent.

The extent of superficial area of that portion of the United States over which the railways have mostly been made, does not exceed one million of square miles. This shows one mile of railway to 38½ square miles of territory.

The following table shows a comparison of the general features of the railway enterprise in this country and great Britain.

	Length in Miles.	Sq. Miles of Territory.	Aggregate Cost.	Sq. Miles of Territory to one mile Rail.	Gross rec'pts per Annum.
Great Britain .	9,000	121,000	$1,500,000,000	13½	$112,584,235
United States..	26,000	1,000,000	1,000,000,000	38½	106,013,600

It appears that with a trifle over three times the territory per mile of railway, the earnings per mile are about one-third in the United States, what they are in Great Britain. The earnings on capital expended for construction in the United States are about 41 per cent. greater than the earnings on capital expended for construction in Great Britain If the latter had only expended on 9,000 miles what the former have expended on 26,000 miles, the British investment would have been satisfactory as to income.

The total area of the United States territory is probably double that given above; but as one-half may be considered either very thinly inhabited or an uncultivated wilderness, to which railways have never been extended, there is no reason to include it in this comparison. The population of the two countries, at this time, is not probably very different, and the gross earnings compare very well in this respect. The British railways are mostly double track, while those of the United States are mostly single track.

As population and business increase, and more railways are wanted in the United States, it will, to a great extent, be accommodated by making double tracks on existing roads, at about one-third the cost of the first track. We cannot expect so dense a population as that of Britain, at least for many years; but it will rapidly increase and become much more dense than it now is, and this will increase traffic. There are thousands of miles of railways in the western States, that traverse districts not more than one-sixth to one-half occupied, an encouraging circumstance for such as are struggling with inadequate traffic; as time, with good management, is sure to improve their income, unless rival lines are constructed to interfere with them, as the growth of the country proceeds. On this point, however, it is to be considered, that as population and business increase, the demand for further railway facilities will stimulate the construction of new lines, and to a greater or less extent, according to circumstances, they will prevent this benefit from flowing to the old lines. It is not probable that capital will seek railways, merely as an investment of funds, for some time to come; but the railway has become so much a necessity, that local interests will make great efforts to do what capitalists hitherto have been willing to do for them. The local interest, when it becomes strong enough to act, will not be so much controlled by the expectation of dividends as by the rise in the value of their "houses and lands," and the general promotion of their business interests.

Men readily become habituated to look with indifference on the most important improvements, and this great work of art, so eminently influencing all the affairs of life, has already come to be viewed as a familiar, if not an ordinary medium of everyday business. In our devotion to the onward pursuits of life, we are prone to seize the present and almost the future, but rarely indulge in the retrospective, or stop to consider the element that so materially aids us in our onward career. Thus, in a measure, we limit and restrict the horizon of our view, and deprive ourselves of the benefit and enjoyment of contemplating the history and progress of those useful arts that have been greatly instrumental in the promotion of our material well-being, and of our progress in intellectual and moral civilization; and it should not be forgotten, that the railway is "an epoch in the affairs of mankind."

The expenditure for this great improvement has been large, and in England and this country has been almost wholly contributed from private sources. The country at large has received benefits that in many districts have doubled and in others quadrupled the cost of the railway, in the increased value of property. Notwithstanding the vast benefit conferred on the public, the average net income to the proprietors has fallen far short of the remuneration to which such outlays are entitled. The net income has not been uniform—in some cases it has been fair, and even liberal—in a more numerous class it has been moderate, and to a large number of stockholders the property has become, or seems likely to become, a total loss. This is certainly a hardship for men whose funds mostly contributed to the construction of works that have been so beneficial to the public. The hardship is more intense when the fact is considered, that these funds have, to a large extent, been furnished by persons in moderate circumstances, by widows and orphans, and by the aged and infirm, who have been induced by shrewd stock-dealers to invest their small properties in railway securities, under the most plausible assurance that they were perfectly safe and would produce a liberal income. To some extent, these representations were, no doubt, honestly made; but this charity can by no means be extended to the whole. In some instances the proprietors have realized their expectations. Railways that have had a fair basis of business, and have been conducted with fair capacity and fidelity, have returned to the proprietary interest a fair, and in some instances even a liberal remuneration for their investment. And instances are not wanting, where lines, regarded in the outset as offering very slender prospects of remuneration, have, by the care, skill, and fidelity of the management, proved to the proprietors a source of steady and reasonable income; while incapacity and dishonesty have rendered unproductive, railways that should have been, from the liberal traffic they commanded, among the best paying institutions of the country.

The celebrated roads of antiquity were made for

military purposes—they were little demanded, and little used for commercial intercourse. The railway was a demand of civilization, and consequently has been appreciated according to the progress in this respect of different people. It is an improvement that in an eminent degree has commended itself to the wants, the interests, and the happiness of men. It was ushered suddenly into notice, and extended with unprecedented rapidity. It opened a new field for the exercise of mechanical ingenuity and the researches of science, demanding investigations and modifications to meet its numerous and hitherto little known requirements. Though the railway had been in operation more than a century, it had been confined to short lines, mostly in mining districts, propelled by horse-power and a few locomotive steam engines, with sundry inefficient means of traction, the most effective of which was the cog-wheel, working in a rack. In this condition it was unable to meet the wants of general commerce, and the labors of several engineers were engaged for years to ascertain some more efficient means of applying steam locomotive power for propulsion.

The attention of the late Mr. Nicholas Wood was directed, for several years, to ascertaining how far the adhesion of the wheel to the rail could be relied on for this purpose; and the world is much indebted to his investigations and experiments. From the imperfection, and want of adaptation of machinery, his results were slow; but they served to direct attention, and inspire confidence that this was the best means that had been proposed, and the practical result on the Stockton and Darlington Railway greatly strengthened this confidence, and with it the idea, that the railway would become an important medium of general commercial intercourse. The opening of the Liverpool and Manchester Railway confirmed the principle of adhesion, and even Mr. George Stephenson found he could depend on this as a substitute for stationary power on his inclined planes, having a rise of a 100 feet per mile. Thus, after slowly groping its way for years, it was discovered that steam locomotive power, by this simple means of adhesion, would be effectually and economically used for the propulsion of carriages on railways. The result is a striking instance of the time and labor often required to demonstrate and bring into use improvements of great public utility, depending on so simple a method of action, that we wonder how it could so long remain undeveloped.

It now remained to consider and devise the best mechanism for the railway, and the various appurtenances required for its complete operation. Very little engineering experience on this subject existed at that time, nor were the general characteristic wants of the improvement appreciated, or how the material and works of the varied mechanisms required, would successfully answer the almost unknown service they were expected to perform. It is therefore no matter of surprise, that many errors were committed—that numerous suggestions of plan and improvement were found inapplicable, and by degrees, as experience demonstrated their futility, were laid aside. In our own country these have been numerous, and even in England it cannot now be said, they have arrived at perfection, or uniformity in the machinery or management of railways. Though they have undergone important changes, their engineers are still divided in practice on important features.

We have had a railway gravely urged on the pannier, or single rail plan, with the load suspended on each side; and this especially commended for timber districts, where it was proposed to cut off trees at a suitable height, and lay the rails from stump to stump as supports for the rail. Singular as this now appears, I knew a board of directors to go to the expense of fitting up a piece of railway and a pair of pannier cars to try the experiment of its working, and their engineer had no small difficulty in persuading them to abandon the plan. The construction of an important railway in this country was commenced, and a section laid and put in operation, with the flanges of the carriage-wheels on the outside of the rails. Locomotive engines were made with joints in their machinery, so as to be borne by two wagons, or trucks, connected together by a movable joint, so that each truck would take its own line on the rail, while the machinery of the engine would move on the curved rail, by means of its movable joints. The anti-friction car was introduced at an early day, and for a time was, with many, a great favorite. All these were presented by men of talent, who had given much attention to the subject. The atmospheric railway had, at one time, a great run of popularity—was advocated by distinguished engineers, and several railways were constructed in England on this plan. These instances are mentioned as reminiscences, not having on trial met the requirements of railways. Others might be mentioned, but these are sufficient to show, that although the great principle of its successful operation had been established, the wants of its varied mechanism were but imperfectly developed.

The history of railways has proved that serious

errors originally existed in regard to the durability of rails and rolling stock. The iron rails were formerly regarded as highly indestructible, and the wear being supposed to be very slight, would answer the service a great number of years. The machinery, in great part of the same material, it was thought would be very little subject to repairs or renewal. For a time, the unexpected deterioration of rails was attributed to light weight; and this was increased until the rails were generally made double, and in some instances about three times the weight of the original rails on the Liverpool and Manchester Railway. At the present time, the tendency is to fall back to a less weight, as best calculated to secure a good article, and consequently a higher durability. The engines and cars have required large repairs, and more frequent renewals, and consequently the expense for repairs and maintenance of stock has been materially greater than anticipated in the early history of railways. This error, no doubt, has arisen, in a great measure, from the failure to appreciate the severity of the service to which they have been exposed.

In the early operations of railways, the locomotive engines weighed from six to ten tons, carrying from one and a-half to two tons on a wheel; the weight has been increased to four and five tons or a wheel. The weight of cars was from three-quarters to one and a quarter tons on a wheel, which has been increased to two and two and a half tons on a wheel. The early speed of passenger-trains was from sixteen to twenty miles per hour—it has been raised to thirty miles and over per hour, and a running speed of forty-five to fifty miles per hour has been not unfrequently made. Some may have anticipated this speed, but not its effect on the rails and machinery. When Mr. George Stephenson was interrogated by the opposing counsel before a committee of Parliament, he claimed that his engine could run ten miles per hour. With a view to lead him into some extravagance, he was asked if "he did not think his iron horse could run fifteen or twenty, or even twenty-five miles per hour." To which Mr. Stephenson, forgetting the caution that had been given him, modestly replied, that he thought it could; an indiscretion, that at that time was regarded so extravagant, as nearly to disconcert the applicants for the railway bill then pending before the parliamentary committee.

Since the time above referred to, there has been great demand for increased speed; and this influence has led to the necessity of more powerful engines, and, consequently, to greater weight. Instead of six to ten tons, their general range has of late been 22 to 28 tons, and some even greater. These weights are exclusive of tender. The running speed has gone up, as before stated, to 50 miles per hour, and a large proportion of trains now range on their running time from 30 to 40 miles per hour. It has been contended by many that high speed was but little more expensive than low speed—a very superficial error. It has also been urged by high engineering authority, that the safety of travelling at 30 miles per hour was no greater than at 100 miles per hour; an equally superficial and inconsiderate error. Whatever may be the importance of high speed, it can only be obtained at a corresponding sacrifice of expense and safety. No intelligent mechanic can witness the passage of a heavy express train, as it thunders past him at a speed of 45 to 50 miles per hour, without at least a momentary wonder that the materials can possibly sustain the rapid action of the parts required under such motion. This branch of the subject will be more particularly discussed hereafter; it is now alluded to for the purpose of showing that the progressive demands of railway service have called for machinery and repairs not originally anticipated; and, hence, much of the original error was a mere want of foresight in providing for the increasing demands that have been made for rail way accommodation. This cannot be regarded as surprising, when we consider the extraordinary revolution the railway has wrought in regard to rapid transport.

It has been stated that the railways of this country have cost, in round numbers, $1,000,000,000. This is not far from ten per cent. of the real and personal property of the United States, as estimated for taxation in the year 1856, and about double the outlay of Great Britain, in proportion to the wealth of the two countries; the former being estimated at $11,000,000,000, and the latter at $30,000,000,000. Ten per cent. of the aggregate wealth of a country, and especially a new country, is a large appropriation for a single means of transport, and that one of a comparatively modern kind, that the world scarcely knew of thirty years ago.

It is not regarded as worth while to set forth in these pages the particulars of the progress and working of railways, the large expenditure called forth by them, and the business results that have attended their unprecedented advancement. More full particulars in regard to these may be obtained from other works already before the public. Having given the general features to show the difficulties—the vast interest of the public—and the large outlay of private funds, I proceed to examine the circumstances of business and the prospect the proprietary interest has to obtain a fair return for the funds appropriated for the promotion of works so eminently advantageous to the public.

CHAPTER I.

CONSTRUCTION.

This branch will be discussed under the several heads that constitute the sources of expenditure.

The first object for consideration in examining a project for a railway, is the nature and extent of the traffic to be provided for. If this is large and of a character to demand high speed, the work must be adapted to bear the contemplated service. Bridges and rails must be stronger than for a lighter traffic and lower speed. If a light traffic, and especially with a lower rate of speed, is anticipated, much may be saved in expense of construction, and also in the expense of operating the railway, by adapting the works to the service to be performed. A careful consideration of this question will have much to do with the value of the investment. This question should be carefully examined before commencing the work, not only to ascertain the nature of the traffic, but its probable sufficiency to warrant the undertaking.

CHAPTER II.

LAND AND LAND DAMAGES.

To a large extent, railways have been excessively burdened in expenses and charges for land. This has been most severe in the older districts of our country. The idea that a man should not be compelled to part with anything he owns, unless he is paid such an equivalent as he may consider himself entitled to, has had great influence on this branch of railway expenditure. The principle cannot be controverted so far as it applies to general objects of commercial intercourse. But railways cannot be made without land, and the land they require must be such as is called for by the line of their works; and they necessarily have no competition, as no other land will answer their purpose. A town or county road is no less dependent on some process of valuation by disinterested parties. If there were no proper way to take it, except on the general principle of private purchase, the means of intercourse, not only by railway, but also by public roads and canals, would be greatly embarrassed; and these indispensable agents of civilization often held in abeyance, if not totally arrested. The law has therefore decided that legislative authority may prescribe the mode of determining the value, as between the owner and the railway corporation. But this has often been done with so jealous a regard to the rights of the owner, that the corporation have generally regarded it for their interest not to resort to it if they could agree with the owner for double or treble the real value; preferring this to the risk of the legal method of appraisal. The determination of this question by a jury of twelve men, as has often been the provision of the statute, is the most objectionable method, especially if a jury is to be called for each particular case. In such cases juries are very likely to enter into sympathy with the owner, and to have very little for what they consider a rich corporation; and moreover the idea of a return in kind, from the owner, when he may sit as a juryman, may have an influence prejudicial to a just and equitable valuation. By the jury process I have known tenfold damages assessed, and the corporation left with no alternative to escape payment. The most just and equitable mode that I have known, is by commissioners appointed by some court of high standing, which select men of character from a different locality, without influence or nomination from either party. Sometimes the parties claim the privilege of presenting names to the court as candidates from which a selection is to be made; but any court, with a proper sense of its duty and responsibility, will not allow this, for the obvious reason that a man placed on a commission of appraisal by the nomination of one of the parties, is most likely, if not certain, to lose his impartiality, in favor of the nominating party. The court in such cases should take the entire responsibilty, and acquit themselves by their sound judgment and impartiality in selecting men for commissioners whose standing will be a guaranty for the competent and impartial performance of their duty.

It rarely happens that a landed estate is not more benefited than injured by a railway passing through it. The exceptions are mostly on the small estates, or building lots, and the disturbance of buildings, when the property may be nearly destroyed; and this constitutes but a very limited proportion of land required for such purposes. It is not often that the appraisers are allowed to take the equitable view of offsetting benefits against damages. But if the corporation could obtain the land at its fair value, the expense would not generally be large, or seriously oppressive.

Railway corporations were formerly made up, mostly from the more enterprising portion of the people, who were willing to risk something for improvement, and it is certainly a hardship for them to incur the hazard of the enterprise and pay in addition large amounts in the shape of land damages to those who are most likely to secure at least an indirect benefit. If any parties are sure to reap advantage from a railway enterprise, it is those who own "houses and lands" on the route; to these the railway cannot fail to give increased value. And yet it is this class that often impede their progress and embarrass their finances. It is curious, but not very flattering, to see the developments of humanity in its manifestations on the question of land damages, especially in cases where it is obvious the landowner would freely give his land, and something more, rather than have the railway fail of construction. But he considers the question settled, that the work will be done, and proceeds to draw from the treasury of the corporation what he can, thinking that a clear windfall for his benefit. The morality of such proceeding may be easily estimated.

The following anecdote illustrates this branch of railway business: A railway corporation being in want of an agent to procure land for their line.

employed a man highly esteemed for his business capability. He was a stage proprietor, who had made his way from the stage-box to the stage-coach office; and having had much experience with men, he entered on his duties with high hopes of success. When, however, he came in to report his proceedings, he did not appear much elated, and remarked that he thought he had seen human nature in all its forms; but he had come to the conclusion that if a man desired to be fully acquainted with human nature, he must be an agent to obtain land for a railway.

The experience of the stage proprietor was not peculiar among land agents of railway corporations. One owner is devoted to literature, or seeks the quiet and retirement of seclusion, and the sight and noise of trains will damage his meditations or repose; another, for the want of a more substantial basis of damage, brings forward proof to show that the natural beauty of his scenery will be marred, and the poetry of his place destroyed by so vulgar a thing as a railway, and a professed artist gravely testifies this will damage his property twenty-five thousand dollars; another is alarmed for the fear of losing his trade, by the facilities the railway will afford distant produce to compete in his market, or give facilities for his customers to supply their wants at other places. Notwithstanding these varied apprehensions, the railway does not fail to enlarge every department of commerce and useful industry, and to increase the "value of houses and lands."

The hurried manner in which railways have generally been planned and prosecuted has very much increased the item of land damages. If a discreet business-man contemplated the construction of a railway on his own account, he would begin by ascertaining the proper location, and then by conditional contracts ascertain what the land would cost. He would hold the decision, not only as to route, but also as to the main question, whether he would make the railway at all, subject to cost of land and other expenses indicated by preliminary surveys and estimates; and if these appeared too heavy for the probable income and other benefit he might anticipate, he would abandon the project. Under this kind of management landholders would look on the project, not as a certainty, but depending materially on cost of construction and prospect of revenue, to be decided, not by an association of men, among whom there may be a strong interest for the railway, without regard to direct profit, but by a single man who will risk the expenditure, only as the income promises to be remunerative. The landowner in such cases realizes, that if his "houses and lands" are to be increased in value by the railway, he must be careful not to put any obstruction in the way, and he will generally be willing to take a fair price, or donate his land so far as required for the object, and if it appear necessary, do something in addition to insure the prosecution of the enterprise.

If a joint-stock company should adopt a similar policy, and only call in a small amount of capital, sufficient for such preliminary proceedings, and hold the enterprise to the same business requirements, they would obtain the same beneficial results. But it often happens, while many individuals subscribe their money to a railway stock merely as an instrument, to be benefited only by direct income, there are others that subscribe to the stock and interest themselves mainly, if not exclusively, to enhance the value of their "houses and lands." This latter class are in great haste to have the railway in operation, as that is the result to benefit them; and whether it is providently or improvidently constructed is not material to their interest; but they are essentially interested in a rapid prosecution and early completion of the railway. To them it is of small moment whether or not the lands are obtained at a large or small price, provided the work is accomplished.

On the item of land damages, and law and legislative expenses, we have not suffered as English railway corporations have for the same object of expenditure. In a work recently published on "European railways," by Colburn and Holley, page 33, it is stated that the average of British railways have expended (for railways completed prior to 1856) for these items $50,000 per mile of railway; of which $43,000 per mile was for land damages. Even the Hudson River Railway, that suffered severely in this respect, cannot compete by a long way with English railways. The same work, page 23, says: "The sums paid by many of the railway corporations for land and compensation are almost fabulous. . . . . It is recorded that the sum paid to one particular individual was so preposterously large that his heir (?) returned the greatest portion of it as conscience money! . . . . Railway corporations have, in the majority of cases, paid from ten to a hundredfold beyond the legitimate value of the properties purchased by them. . . . . More recently landowners are less clamorous for compensation." It is indeed difficult to get clear of the idea that the above statements are quite "fabulous." Mr. Robert Stephenson, from whom the above writer quotes, says, "almost fabulous." Looking at this as applicable to more than 6,000 miles of railway, mostly through farming lands, and a greater or less proportion of the land of little real value, it certainly cannot be regarded as complimentary to the intelligence and morality of the British nation. Perhaps, had our ideas of the sacred character of landed property been similar in degree to theirs, we should not have fared better; for according to the value of lands, we have practised on the subject with great skill.

CHAPTER III.

LOCATION OF RAILWAY.

The first field duty is that of survey for location. The terminal points are usually determined in the outset of the project, and the work required is to settle the intermediate line, which should devolve on the engineer; at least he should make all necessary surveys, maps and estimates of expense, and lay the whole before the Board of Directors, with his opinion as to the most elegible route. In the early history of our railways this was the usual, if not the general practice. This method places a direct responsibility on the engineer, who is best qualified to meet it, and whose professional reputation is materially concerned to guard against errors that a future observer might discover. On his skill and fidelity much must depend, and the directors of a railway company will need good judgment in selecting one whose experience and skill will be a guaranty that this duty will be discharged with ability and fidelity to the interest of the enterprise. The desirable result, it is evident, has not always been secured. Locations have been made that required large expenditures to correct, or when not corrected have worked a permanent detriment to the line. This has sometimes resulted from incompetent engineering, but not always. Scheming directors and agents, and sometimes speculating engineers have had other objects than the interest of the corporation to control their action, and lead to improper location. When persons of this class obtain a seat in a board of directors, they will be most likely to employ an engineer who they regard best calculated to serve their object; and these are not engineers of high standing for professional skill, firmness and fidelity. Another source of error in this respect is, that of giving the contractor the control over the location within certain restrictions of grade and curvature, by which he may travel a good deal out of the proper location, very much to the permanent injury of the proprietors. It not unfrequently happens that the citizens of a town, in order to secure facilities for their accommodation, are very active and contrive to exert an influence on questions of location, that involve large future expenditures to correct the errors they induce.

In a new country, with its business undeveloped, towns are in some instances started, make considerable progress, and finally prove not to be in the proper line of the commerce the railway is constructed to accommodate, and after falling into this error, the corporation find themselves compelled to change a considerable part of their railway or have their traffic divided by a rival line. In whatever way it may happen, an injudicious location will work a damage to the proprietors.

If the termini of a railway be well settled, the line should generally be located on the most direct route between them, that the formation of the country will permit. It rarely happens that any local object of traffic will warrant material elongation of the line, especially in a new country. A railway (unless for some special object) can only be supported by general traffic, and it is manifestly important, that this should not be burdened for a local object, unless that be of very considerable magnitude; for it is obvious, that so far as it travels out of a proper line, it is at the expense of its main traffic. A further source of error in this respect is, the haste very often indulged, to commence the work, not allowing the engineer sufficient time to perfect his surveys and examinations, and thereby compelling him to act in this important duty, on inadequate knowledge of matters affecting the question. Nothing can be more impolitic or unwise than this, for men who propose an investment of funds. It should be kept steadily in mind that a railway is an expensive thing at best, and to increase this by hasty and inconsiderate action on the important matter of location, may greatly impair the value of the investment, while it only gratifies an excessive ardor, by no means consistent with a wise and energetic pursuit of important business.

The duty of location is a very important one to the proprietors and the public; both being inter ested in effecting the cheapest transport. In pro ceeding upon it, the first object should be to under stand the general line of the trade of the country through which the railway is to be made. Rail ways that cross this line, at or near right angles, are seldom successful. If no error be committed in this respect, what remains is essentially to obtain a thorough knowledge of the face of the country between the termini, from which the relative advantages may be reduced by calculation and estimates of their bearing on the cost and the commercial importance of intermediate objects. On some lines the physical formation of the country clearly points out the proper route. Others require extensive surveys to determine it. The latter are most likely to suffer for want of competent engineering. Instances have occurred in which a considerable amount of new railway had to be made, to correct errors of location, and receive the traffic the line was originally designed to accommodate; thus loading the enterprise with large additional capital, that would not have been necessary had the original location been correct. But though the error may not be such as to lead to the making of a new line for any part, it may prove a permanent injury by loading the traffic with unnecessary current expense. From whatever cause this may have happened, it is equivalent to a loss of capital, and consequent diminution of dividends, necessarily impairing the proprietary interest, especially that of stockholders in the institution.

After the settlement of the terminal, and (if there be any) any other controlling points on the route, it is important to secure the most favorable grades and allignment. On these points little need be said; it being obvious these should be as easy as the formation of the country may permit within the range of suitable expense of construction. It has sometimes been the practice to adopt a uniform curve for all points that produce angles in the line, and in others a maximum and minimum rule has been adopted. The best method is to adopt the largest radii, or least practicable curve the formation of the ground will permit. The mode of construction of railway carriages is adapted to run on straight lines, and any departure from this must be forced against additional friction. Sharp curvature is also objectionable, as obstructing the view of approaching trains.

In a line having considerable preponderance of trade in one direction, there will be advantage in inclination in this direction, corresponding to the preponderance in the weight of traffic; otherwise a level is the best grade for a railway. This cannot always be obtained, and indeed on most lines, only to a partial extent. The practice should aim at the least practical departure from it. To reduce what is termed the ruling grade, will often justify pretty heavy expense, and more or less elongation of line, as this feature will control the weight of trains.

CHAPTER IV.

METHOD OF BUSINESS.

The method of business pursued in conducting the work of construction has been widely different at different times and on different railways. The plan adopted in the commencement of railways was to have maps and profiles of the line, and plans and specifications of the manner in which the work was to be done. As soon as these were prepared, the work was advertised for contract, and let to the lowest bidder who was considered responsible for the undertaking. Propositions for the work, received in this way, were reduced to contracts, providing for payment at certain rates on the several items. The engineer was made the inspector and the umpire between the parties, from whose decision there was no appeal in regard to anything pertaining to the contract—the manner of performing the work, the measurement and estimate of quantities provided for in the contract, and the valuation of any extra work that unforeseen circumstances might call for in the course of construction. The same mode of proceeding had generally been adopted in the construction of canals. In this method, the engineer stood between the corporation and the contractor; and upon his capacity for his duties and fidelity to the parties, the system very much depended. A want of confidence could not fail to produce dissatisfaction, and it is obvious it could only be maintained by the administration of engineers of sound business experience and unquestioned fidelity of character.

In the early history of railways, the line was divided for construction contracts into sections of one or two miles in length, and contracts made for the grading of each section; and for light work several sections were sometimes let in one contract. But the large amount of this kind of work soon called into this service contractors whose ambition aspired to larger contracts, and they gradually drove from the ranks of original contractors many that had been content with moderate contracts, and who depended on their small capitals, and a personal supervision of their work, for their remuneration. The latter were either compelled to abandon their occupation or employ themselves on subcontracts from their more influential associates. Eventually, contractors did not feel satisfied with anything short of the entire railway; including grading, bridging and laying track, with some station buildings and rolling stock, with the addi tion in some instances of the iron rails, chairs, spikes, etc. So long as the original principle was maintained, and the contract made with carefully prepared specifications, and a competent engineer placed in full authority as the inspector and umpire between the parties in the contract, and the compensation determined by the measurement of each item of work done, as originally practised, the evil of large contracts was not so great. Still it was not the best method. The advantage over it of several contractors, is, that work is generally better done at lower rates; and as contractors sometimes fail to perform their contracts, either by doing their work improperly, or not as rapidly as their contracts require, and the work has to be re-let, in such cases when there are a number of contractors on the work, the failure of one is of less importance, and involves less difficulty and delay in re-letting than when all, or the principal part of the work is in the hands of one firm. A further difficulty occurs most usually from letting in large contracts to one firm, from the disposition to sub-let the work, and thereby introduce upon the line men as contractors whom the officers of the railway company would not accept, often causing embarrassment in the prosecution of the work. If, as is usually the case, the contractor is restricted in his contract in regard to sub-letting, he will more or less contrive to evade or disregard the restriction.

For some years past it has come to be a very common practice to put the entire work of constructing a railway into one contract. The influence that has produced this practice has generally impaired the authority and power of the engineer, which was naturally to be expected, as the contractor who can wield an influence to control so much work is not likely to be controlled by an engineer. He looks on the engineer as only useful to set stakes and levels, and perhaps make estimates for his sub-contractors. In some instances the engineer has been appointed subject to the approval of the contractor, and subordinate engineers employed and paid by the contractor directly.

One of the most common methods adopted by contractors to obtain the entire work of a line, is to contract for a large part of their pay in the stock and bonds of the company. It will readily be perceived that parties desirous of securing the indirect benefits of a railway, and not able to obtain sufficient cash stock to do the work, would readily fall into such a method of contracting. And if such parties (as is most probable) have the controlling management of a railway enterprise, they will find plausible reasons to satisfy what cash subscribers they may chance to have, that it is the wisest policy to contract with one firm for the entire work; that they may know in the outset what the work will cost, and not be subject to the uncertainty of an engineer's estimate. This is a plausible fallacy, as experience has abundantly demonstrated.

Now, if the object be to secure the incidental benefits that may result from a railway, this method of contracting may often be the best, if not the only one practicable; and if there are no stockholders, except those who are so interested in the indirect benefits of the railway that they regard income on their stock as a secondary object, this plan of contracting may be very well. Under these circumstances, a subscription is got up among parties so interested, with a view of raising a portion of the funds required, and mainly to give currency to the bonds of the company that must be relied on to a large extent. To do this, it is necessary in most cases to obtain subscriptions from municipal corporations of towns and counties; and so a very respectable amount of subscription to the stock is obtained, and the work goes on. This is all very well so long as the original parties only are interested. If, however, a cash subscriber falls in, who has only interest in the income that pays dividends, he will most probably be disappointed. To raise some funds on stock, and give apparent basis for bonds, the plan may be presented for illustration as follows: A limited amount of stock will be taken by individuals having a strong indirect interest in the business of the railway. This subscription will, to a greater or less extent, fail to be paid. The municipal subscription will come in the form of town and county bonds, guaranteed by the railway company; these bonds will be sold at a discount according to the credit that may be given them; the private subscription may in part (as frequently done) be paid in farm mortgages, guaranteed by the railway company, and sold by them in the same manner as the municipal bonds, at considerable discount. These are the cash elements on which the contractor begins his work. To obtain the iron rails, it is common to issue first mortgage bonds to a limited amount. The contractor must, of course, take a large part of his pay in the stock and bonds of the company. It is apparent that the contractor, in regulating his compensation, must take into account the value of the material he is to receive in payment; and this will depend much on his ability and skill in financiering his stock and bonds. If the enterprise be faithfully carried out on this plan, it is evident the nominal capital will be much above what would be required in constructing a railway for cash. But the method under consideration has further and peculiar tendency to swell the cost. In such contracts, by a single firm for an entire railway, it is usual to provide for only so much work as will be indispensable to run trains; ballasting is rarely provided for to any further extent than may be furnished by materials from excavation of prism; and the result is, that the work contracted for is not much more than half done, and consequently requires large expenditures after it is accepted by the railway company, showing conclusively that the contract did not determine the cost of the work. The evil that has resulted to railway property by this course of business has been disastrous according to the necessities of the enterprise, the character of the managers, and the degree to which the system has been carried. Under the most favorable circumstances, the wholesale system of contracting, even on the cash plan, enhances the cost of the work, from the causes before mentioned. Its tendency is to undermine the authority and usefulness of the engineer as inspector and umpire, if it does not entirely override and dispense with his authority, and inevitably gives the railway company work inferior to what they contracted for, and compels them to materially increase their capital, in order to bring their work to the standard required by the contract. In some contracts the contractor has been allowed to vary the grades and curves within certain limits. Hence it was his interest (as his contract is usually for a round sum) to make many modifications in these respects, that prove a permanent source of embarrassment and injury in operating and conducting the traffic of the railway, necessarily impairing the interest of the proprietors; and, instead of the contract settling the cost of the work, it leaves the question, even after it is opened for use, totally unsettled, except that the cost is to be looked for much beyond the limits of the contract, and represented in second and third mortgage bonds, under which the value of the stock gradually depreciates, and sometimes passes out of view. Though I regard the mode as highly objectionable, it must be admitted a few railways constructed in this way have turned out to be fair investments for the stockholders: such have been located in a comparatively new country, with great facility for cheap construction; strong natural sources of traffic, that were rapidly developed by the railway; and having fallen into the hands of favorable managers, the evils of the system were only partially experienced, and the traffic proved sufficient to afford fair dividends. This, however, does not prove the system good; it is only an exception to a rule generally bad. The railways more recently constructed on this plan have added to their capital from 25 to 100 per cent. over the amount that would have been required by a sound system of business; and hence one cause of the depreciation of this kind of property; and there can be no doubt the old method of business is the best for parties who furnish cash to build railways, as an investment of funds.

Amalgamations have been made, and branch lines constructed, under the plausible idea of increasing the income of the main line, while the result has shown that, in most cases, the capital has been increased proportionably more than the net income, and the only party benefited has been some scheming contractor or jobbing manager.

The financial business of railway corporations was originally commenced by opening books for subscription to the stock; and having secured the amount supposed to be sufficient for the undertaking, installments were called in as wanted for the work. In this way the principal capital was raised. It was regarded necessary to have at least half, and sometimes three-fourths or the whole of the amoun needed in the form of stock. If bonds were issued it was not done until the investment by stock had made a good basis for the security of the bonds. The bonds were usually sold under advertisement to the highest bidder, provided there was no restriction as to usury, and commanded par, or nearly par, for seven per cent. bonds.

Gradually the old method of conducting the finances gave way to a different system. In the ardor of the railway movement, opportunity was afforded for men to enter into their management, not faithfully to discharge a trust, but for the purpose of enriching themselves at the expense of those who confided in their supposed fidelity. Such men have ruined the stock, and greatly impaired the interest of the bond proprietors of railways that should have been the best investments of this kind in the country. A small clique of such men will often baffle the efforts of a majority of fair-minded directors, and gradually deplete the proprietary interest. They enter the management with fair pretensions, and contrive to obtain sufficient, if not most of the proxies for elections. Where the railway is far distant from most of the proprietors, and they know little except from reports, the means for controlling the institution is not difficult to compass. It is hardly worth while to go into particulars, at this time, of the process of the financiering of such men; they have left their marks too indelibly fixed on the railways that have come under their financial management, to require anything more than an allusion to this source of depletion, to which railway proprietors may look for a large portion of the loss of their property. This, with the method of contracting, the construction of branch lines, and amalgamations, has so overloaded many of our railways with capital, that the prospect of the stock proprietors is very discouraging, if not hopeless; and even the bond proprietors find their interest greatly depressed, and what they had regarded as very safe, has proved to be very uncertain.

There can be no doubt the old method of finance and contracting was the best. It had sound business principles to rest on, and by it, enterprises of a difficult and expensive nature have been carried through, and succeeded as fairly remunerating investments of capital. A board of directors, with fair business capacity and integrity of purpose, with a competent and faithful engineer, will accomplish all that can be obtained, and if they do not make the enterprise profitable to the proprietors, it will be for the want of sufficient inherent elements of success in the enterprise itself; and no system of wholesale contracting and jobbing financiering will do as well, if investment of capital be the object of the enterprise.

A practice has prevailed, I believe, very generally in England, of jobbing the engineering of railways at a percentage of five or more per cent. on the cost of the work; the engineer employing and maintaining such force in this department as he judges expedient. There are two serious objections to this method: First, it offers an inducement to the engineer to increase the cost of the work—and secondly, to employ an inadequate force and capacity of assistants to meet the proper supervision of the duties that devolve on this department of service. It encourages a speculating spirit of action in the engineer; and this once getting hold of his mind, will be very likely to lead to many abuses. A fixed salary should be the only pecuniary resource an engineer should have in view, in a work committed to his supervision, and if he possesses the morality that should control him, his entire energies will be devoted to the duties he owes to the proprietors, and as umpire between them and the contractors. An engineer devoted to the acquisition of knowledge, in the noble and responsible objects of his profession, and the attainment of the high character that should animate and possess those who are charged with the care and management of the interests of others, will have no time for speculating schemes; and it may be set down as a rule, that an engineer given to speculation and effort for a rapid accumulation of wealth, will be superficial in his professional acquirements, and very unsafe in his duties generally, and especially as an umpire between the proprietors and the contractors.

It is often remarked, that the estimates of engineers, made of the cost of projected works, are much below the actual expenditure. No doubt this is very generally true. The engineer who makes the estimate must bear the responsibility, at least so far as professional reputation is concerned. He is exposed to the suspicion of acting with a view to obtain occupation in the prospective supervision of the work. He is not, however, the only erring party. Men who have an indirect interest to serve, are often engaged as promoters, and perhaps managers of the preliminary business of the enterprise; to whom the actual cost and net income are secondary considerations—subordinate to the incidental benefits they expect to derive from a railway. Their object is to induce capitalists to furnish funds for construction; and to make it as promising as possible, a low estimate of cost is advantageous.

Such persons accompany the engineer in his examinations, and usually represent the facilities of the work, in the most favorable light. All kinds of material wanted for construction are easily and cheaply to be had; and if some desirable kind does not present itself, substitutes, though of acknowledged inferiority, are strongly urged as proper and suitable under the circumstances—the aspects of trade are urged as offering the prospect of low rates for all kinds of labor. These are pressed on the consideration of the engineer, not merely by an individual, but generally by the influential men he meets in the course of his examination. To meet these influences, he has comparatively small means and limited time. Often the parties that bring these influences to bear, furnish the means for the preliminary survey and estimate, and the engineer is not allowed to be ignorant of the result of an unfavorable estimate; namely, that the money furnished for the survey will be lost to those who have been his patrons. To all this it will be properly said, the engineer should not allow himself to be controlled by influences that lead him into professional errors. A less reprehensible source of error, in original estimates, is found in the inadequacy of the views and plans of work, to meet the unexpected demands of the traffic: the developments of business raising this sometimes 100 per cent. above the original estimates of trade, and consequently demanding works of more capacity of accommodation, and necessarily of greater expense. This state of things is not unusual, and while the cost of construction is carried up to meet a larger traffic than was anticipated, it is not reasonable to hold the engineer responsible for the excess that arises from this contingency.

The duty involved in the preliminary survey and estimate for an important railway enterprise is no easy task. To do it properly, so as to secure a reasonable approximation to the cost, the engineer should have sufficient time and means to make surveys, essentially adequate for a location of line—by trial pits obtain at least a fair general knowledge of the soils to be excavated—prepare plans for the various structures in sufficient detail, to enable him to make computation of the quantities for the several kinds of work required—and to ascertain the existing facilities for obtaining materials for the wants of the various structures necessary in the work. Thus prepared, an estimate may be made that would form a reasonably safe basis for the financial interest involved in the construction of a railway. This method, however, is rarely adopted. No one perhaps will object to it, as being a sound and sensible mode of reaching the object. But it requires considerable expense and time to obtain the desired information in this way; the means may be wanting to meet the expense, and patience is wanting to grant the necessary time.

If a company of discreet business men should undertake the construction of a railway for the single object of making it an investment of funds, they would proceed in the manner last described, preferring to risk a loss of one per cent. in obtaining reliable knowledge of the probable cost, to that of hazarding greater loss by trusting to insufficient examination. It does not invalidate the argument, that the proceedings in the preliminary steps for this kind of enterprise have generally been of a different character; and that, notwithstanding this deficiency in examination, they have sometimes been successful; for while cases of success have occurred, many have resulted unfavorably, greatly disappointing the expectations of the promoters of the enterprise.

There are two legitimate objects that appertain to the construction of a railway, namely, the direct one of making it a profitable investment as a dividend-paying enterprise; the other, the indirect benefits to trade, general intercouse, and to increase the "value of houses and lands." For either of these, a man may legitimately seek to promote his interest by furnishing funds for a railway; and in either case, he will need the same care in all the methods of business. Other inducements to promote the enterprise may be found by speculating men, who contemplate benefits to arise out of the current business of the corporation—as the contracting, that looks only to the chance of a good job, to be secured by a position of influence; also the financiering, that looks to the use that may be made of transactions in the stock, bonds, branch lines, amalgamations, and other fiscal operations of the company—and have no further regard to the funds, or the economy of the work, than may be necessary to secure their personal object. However little desirable, the latter have been fully represented in many important enterprises of this kind, as the history of railway management too clearly shows. In England they had one of this sort, who was styled "railway king," and at one time many people supposed it hardly possible to build a railway without the support of his coöperation. He was regarded a man of extraordinary capacity, and his advice was sought as of one possessed of incomprehensible powers. The deception, however, as to his great powers of business, was not of long continuance. His frauds were discovered; the diadem stripped from his brow, and his seat in parliament did not secure him from the degradation so fully earned by the abuse of misplaced confidence. Our country has not escaped the influence of this class of men, nor can we claim exemption from the blind devotion that gave them power of mischief. In proportion to capital, they have probably been as injurious to railway investments in this country as in England. If their history could be read by railway proprietors, it would point out a cause for a large share of their losses, which arise from excessive capital in the construction account, and which weigh down the stock of many railway corporations with a load that hardly admits the hope of recuperation; and this on lines that, under sound management, should have been good investments of funds.

From what has been said in relation to contracting, no imputation must be inferred in respect to the honest contractor, who in fair and open competition takes his contract; nor are the contracting parties to be regarded as so mischievous as those who hold indirect interest in their contracts by means of official influence. The great field of action for depleting a railway corporation will be found, for the most part, within the board of directors; at the same time, there may be a majority, or a respectable minority of intelligent and fair-minded directors, who would do nothing knowingly to injure the property of the corporation. Two, three, or four men in the direction generally conduct affairs; and if these are fair and upright men, the institution may be properly conducted. But if not, they may easily manage affairs with reference to their personal interest, which will be developed by placing men in their own interest in all positions of trust, and keeping the books to their own liking, so that other directors will find difficulty in obtaining information of the real course of proceeding And if perchance any one becomes suspicious and troublesome, no effort will be spared to impair his influence. He is impracticable, an old fogy, not up to railway times, or is influenced by jealousy; and his less suspicious associates, busied with their private affairs, and having no time for long stories, pass over his complaints, and put them down as the result of a suspicious mind, and no serious objection is made at the next election to leaving out of the board the impracticable director. Now, perhaps with a little more caution, the ruling men go on, more or less to the damage of the proprietors. No doubt much of the evil of unfaithful management arises from the fact, that directors have generally but little practical knowledge of railway business, and consequently are easily misled by artful and unfaithful associates. The business, moreover, is of great detail, and requires much time to obtain an intelligent knowledge of affairs, especially if not conducted on the most direct and strict principles of business; and directors are not generally disposed to give the time necessary for thorough examination. They receive little or no compensation; their position is regarded an honorary one, and they naturally expect the officers to do the work; if they act on a committee, they expect all the work prepared to their hand. This is clearly wrong; they should make full investigation and know from original sources the basis of their action. But this would require much time, and the officers, very likely, would feel themselves insulted by such manifestation of want of confidence, and if countenanced by their friends in the direction, they might, by indirect proceedings, greatly embarrass, if not defeat the object of the committee. Such interference, however, if attempted, would cease, if the practice of investigation was vigorously maintained. A committee of the board of directors, under no circumstances, are of service to the proprietors, if they are of that party in the board that may hold rule adversely to the proprietary interest, as they are to be regarded a mere cloak to more effectually hide what should be seen. It is apparent, the safety of the property must depend on the efficient action of a competent and faithful board of directors, and it would be better for the interest of the institution to make them such compensation as would be reasonable for the time devoted to the management, and thereby fix a business responsibility; to ask them to devote their time, and to neglect other business, without remuneration, is not reasonable, and rarely expedient. Very few men devote themselves to transact the business of others, without compensation in some form, and if it is not direct and open, it may be indirect, and this no-compensation may be regarded as liable to be of the most expensive kind for the proprietors. Honor will satisfy some men that can afford the time, but, in general, the ox that treadeth out the corn, must needs have a portion.

CHAPTER V.

GRADING.

Grading is understood to embrace all the work required to bring the surface of the ground to the grade lines, and is mostly earth-work; for a limited extent of line it is rock-work. This work prepares the bed for the superstructure, and when the earth is not composed of sufficiently hard gravel (and this only happens occasionally) it is excavated to a proper depth, so as to give space for the ballast. In this branch of construction the drainage is to be provided for, and is a very important part of the work. The most experienced engineers were early impressed with the necessity of thorough attention to this, as essential to a good railway. In soils through which water percolates freely, as coarse gravel and sand, the drainage may be so well provided for as to require little attention; but most soils are too retentive for this, and require drains to be so made, as to take the water off quickly to some natural or artificial channel, which will carry it beyond the reach of influence on the road-bed. It is not sufficient to reduce the level of the water merely to the base of the road-bed, or foundation of the ballast. Water standing at or near this level, will soften the foundation and allow the ballast to settle, and thus derange the superstructure. Standing water at this level, while it leaves the top of the ballast dry, will, by capillary attraction, rise so as to soften the bottom, and particularly the close retentive soil that forms the bed of the ballast. An embankment that has a standing pool of water against it, though three or four feet below its top, will be impaired in its stability for supporting the superstructure, so as to require frequent adjustment of the rails. Whether from rain or springs, the water should have quick passage to some point beyond the reach of influence on the road-bed ; and if the embankment be exposed to standing water that cannot be removed, its height, if practicable, should be raised, with heavy gravel and stones, at least four feet (and six would be preferable) above its standing level. It is not often this difficulty exists to an extent that cannot be removed by opening ditches so as to lead the water to lower ground.

It is not often that much difficulty or expense is encountered to protect embankments against the influence of water; they are raised above the natu ral surface of the ground, and by moderate ditching their base is easily laid dry. According to the nature of the soil on which they rest, they require more or less attention, in order to insure their stability, and give regular support to the superstructure. Where circumstances give a choice of grade, it is always best to adopt that which raises the rail above the natural surface two or three feet; this with small side ditches generally gives a good drainage: it makes a cheap road-bed, and is particularly important if ballasting material is not convenient. It generally happens, however, that such locations admit of cheaper grading, by running the grade nearer the natural surface of the ground; a circumstance that a contractor working by the mile or under a lumping contract, who has power to modify grades, will most likely take advantage of, to the detriment of the proprietors.

In excavation, there is no way of obtaining adequate drainage but by side ditches, that are low enough to take the water from the road-bed. This requires the width of excavation beyond what is required for the road-bed, to be sufficient for the side-ditches. The bottom of these ditches should be not less than six inches below the foundation of the ballast, when the latter is two feet deep, or with more shallow depth of ballast, not less than two and a half feet below the top of the ballast, and provided with a quick discharge, that will readily carry off all falling water. If the ground be springy, and furnish water constantly, or the greater part of the time, the ditches should be deeper, so as to effectually take the water from the foundation or road-bed. The ground that forms the foundation of the ballast should be shaped so as to easily lead the water that settles though the ballast to the side ditches. In long cuttings, with level or nearly level grades, it is advisable, where circumstances permit, to raise the grade at the point from which the water will most readily flow, so as to give more free and easy discharge from the ditches through the cut. The inconvenience to the traffic in such cases, by the extra rise of grade, will be more than compensated by the superior condition in which the track will be maintained under more perfect drainage. It is obvious different soils will require different degrees of attention to drainage; but it is not probable any will receive more attention than will be useful. To make the necessary ditches in deep cuttings involves considerable expense in opening the cuttings wide enough to provide for the widths and slopes of the side ditches. From the desire to avoid as much as possible this expense, the ditches are generally inadequate; small at first, and soon filled up from the wash of the banks. The width required for a side ditch and its slopes will vary in different soils; but it will rarely happen in soils of a retentive kind that less than eight feet will be sufficient to do the work well. This for the two ditches requires sixteen feet to be added to the width of the cutting, besides what is necessary for the road-bed. This extra width of cutting, it will be perceived, does not increase the excavation in proportion to the whole cut; for the slopes are required for any width, and the addition is a parallelogram of the extra width and depth of the cutting. If, however, the cut be deep, it will add considerably to the excavation, and hence twenty feet width has often been adopted as the rule instead of twenty-six to thirty feet. In rock-cutting, as no slope is required, the ditch will need much less width; and the drainage is not as important as in earth cuttings. The ballast cannot sink in rock, and if it be of a hard gravel or broken stone it will not suffer much from water; at the same time, it will not answer to neglect drainage even in rock-cuttings, as frost would operate in cold weather on the ballast, and by heaving derange the superstructure: also, if the ballast is not of a superior quality, drainage is necessary to enable it to bear the service of trains.

This is a department of work very likely to be

imperfectly done, if the funds provided are inade quate to the work, and the ruling effort is to get a railway on which trains may be run at some rate. The railways made under general contracts almost invariably fail in securing good, or even tolerable drainage. The after-work that follows imperfection in this respect, usually, and it may be said always, is done at a great increase of cost; and it will generally be a long time before it will be brought to a reasonable degree of perfection. In the meantime great injury is experienced to the rails and rolling machinery. Railways intended to be run more or less (as is often the case) without ballast, or with light and imperfect ballast, are especially in need of good drainage. It cannot be recommended to open a railway for transportation without ballast, but if circumstances demand this, the drainage should be thoroughly done, in order to obtain the best condition for poor materials that is practicable to sustain the service of trains. By thorough attention to this department of work, inferior soil may sustain a considerable traffic. In any case, where there is a want of suitable material for ballast, or if the funds are not at command to put it on, the machinery of the railway should be much lighter than may be suitable for use on a well ballasted railway.

There can be no doubt the railways of our country are greatly deficient in drainage. This causes a large additional expense in repairs of track, and in the wear and tear of rolling machinery. No railway management that neglects thorough drainage can be regarded as sound in economy; the saving in the original outlay will be much more than absorbed by the extra subsequent expenditure—probably three or four fold. If the grade line of a railway has been laid so low that the water cannot be fully taken off by drainage, the grade should be raised so as to carry it out of the reach of this destructive influence.

In a work recently published by Colburn and Holley, on European railways, there are some very appropriate remarks on the subject of drainage, contrasting foreign and American railways in this respect. They go to show the superiority of the English practice over the American.* Drainage is

* Though I approve of Messrs. Colburn and Holley's remarks on the subject of drainage, I must protest against some of their statements of facts. They state the width of cutting on the Hudson River Railway at 16 feet for single and 26 feet for double track; whereas, the fact is, 26 feet for single and 40 feet for double track in earth-work, exclusive of side slopes. I notice other errors in relation to that railway. I refer to the section between New York and Poughkeepsie, which was constructed under my supervision, but do not know the dimensions of this work between Poughkeepsie and Albany.

a feature in this kind of work that should never be lost sight of by the engineer, whether he is con structing new work or supervising a railway in operation.

The embankments or fillings required in grounds that are below the grade level are not usually difficult works, as to the mode of execution. Where the level of the natural surface is not more than five feet below the grade level, the perishable matter should be removed from the base of the bank. For greater heights it is often expedient to remove the coarser kinds, and especially such as would injure the early consolidation of the embankment. In low banks, from one to four feet in height, and all that may require side drains, the first material should be taken from the side ditches, so far as they may afford it of suitable quality, or that sufficiently free from vegetable matter. By judicious management in such cases, the most effective drainage may be secured at small expense. If the ordinary width of ditches does not furnish sufficient material for the embankments, they should be laid out and cut such additional widths as may be necessary, keeping an eye to the most efficient drainage. The extent of this will be controlled by circumstances; as a surplus from cuttings, or more convenient source for borrowing, where no more width need be given tho side ditches than may be required for good draining. So far as the surface of the country will permit the grade line to be established at two or three feet above the natural surface of the ground, it is the most favorable for the construction of a railway; and in laying out the work in such cases, care should be given to cut the material from side ditches, or otherwise borrowed for filling, in a regular form, extending over no more ground than is necessary to obtain the amount wanted. The slovenly, irregular manner in which this is often done, taking unnecessary ground, and leaving bars and patches that injure the drainage and give a barbarous, untasteful appearance to the work and the country, is no credit to the taste, judgment, economy or management of the supervising engineer, if there be any such, not controlled by the contractor. The first resource for material for filling after the side ditches required for drainage are exhausted, should be the excavation from the prism of cuttings. It should be the rule in establishing grades, to absorb the excavations in the embankments; and it will be best, as far as circumstances permit, to provide that the excavations be rather short, and if necessary to have more material from the cuttings, to obtain the balance by an increased width, and thereby improve the facilities of drainage. So far as this method can be carried out, it gives the most neat and finished aspect to the work, avoids unsightly spoil-banks, gives the least slope-face to be subject to the wash of rains, and in every way gives the greatest advantage for easy and effective drainage. If the undulations of the natural surface be short, it is not often difficult for the engineer to establish the grade, so as substantially to secure this result, without unreasonable haulage. But if the undulations be long and rather heavy, the haulage may be long and the expense too great, and spoil-banks must be formed. It will be better to pay some additional haulage, rather than put material into spoil-banks; and the engineer should carefully consider the question. By suitable provision of temporary track, where the quantity of earth to be moved is large, the expense of moving may be greatly reduced, and the cost of moving it horizontally a considerable distance will be less than to put it into spoil-bank, and in most cases it will be found the best economy to encounter the haulage, and put the material into embankments. If the side ditches and excavations from the prism of cuttings do not furnish sufficient material for the embankments, it must be made up by what is termed borrowing: *i. e.* taken from grounds outside of the regular work. This should be done with as much care in reference to form as circumstances may permit, so as to do no unnecessary damage to the lands by waste or disfiguration. By a little attention to this, much dissatisfaction to landowners will be avoided, and neatness in the work secured, without sacrifice of sound economy.

The best mode of building an embankment, in order to secure early and regular consolidation, is to carry the material on by carts and wagons, laying it over the surface in regular courses of six to twelve inches. If the material be lumpy (as is very often the case), the feet of the horses and wheels of the vehicles will break it down and compact it, and thereby secure an even and early consolidation. Where the material is within carting distance, this course can be easily adopted, and is of great advantage in securing early stability to the road-bed. If it be necessary to haul the material by a railway, this cannot be done; as it must be dumped from the cars standing on the grade levels and forms a looser bank than any other mode, not excepting that by the barrow, for in this case, though not effecting consolidation in the process, the material may be laid on in regular courses, and by exposure to the air and rains acquires more compactness and regularity of firmness in the progress of the work, than that done by cars. It is apparent that according to the mode of construction and time for consolidation during the progress of the work, allowance should be made in the height and top width of the bank, to provide for settling. This can only be done on the judgment of the engineer, whose experience must be the guide, as to how much will be necessary to secure the proper height and width after consolidation. In addition to the waste by shrinkage or settlement, the banks will lose by the wash of heavy rains, until by compactness and a covering of vegetation they may be able to resist this influence. For as in cuttings, insufficient dimensions in the banks will cause much after-work, which is always done at increased cost after the railway is put in operation.

The slopes of cuttings and embankments may vary in different soils, but rarely should be less than one and a half horizontal to one vertical, unless there be walling or other artificial means to support them.

Embankments on soft marshy grounds require much attention. If the formation of the adjacent country permit it, the grade should be brought as near the level of the marsh as will admit suitable drainage. If the marsh can be drained only a few feet, it will improve its capacity to bear the railway trains. The object of reducing the grade in such cases is to avoid the weight of bank, leaving as near as possible its total strength for the support of trains. After such improvement by draining as may be practicable, it will usually be necessary to do something to form a broad base for the support of the road-bed. Fascines, or small sapling trees, from half an inch to one inch in diameter, made up in bundles of six to eight inches in diameter, and eight to ten feet long, have been regarded as the best means of securing a broad and equal-bearing base. The mode of laying them is across the track, breaking joints and compacting them together, so as to cover a width corresponding to the nature of the marsh—usually about sixteen feet for single track. If well put in, they form a mass that in most cases is of sufficient strength; they fit well and uniformly on the marshy soil, and being elastic, are better than a foundation of more rigid material. Other methods may answer a useful purpose, as a course of trees, five to ten inches in diameter, laid across the track, and closely chincked by a smaller and shorter course, under the road bed. Courses of plank may be put down, taking care to have an extra course of shorter plank under the road bed, so as to diffuse the pressure on the course first laid. Longitudinal sleepers for a support to the cross sleepers of the track may be beneficially adopted in some cases, to further diffuse the pressure on the foundation. It must be kept in mind that any plan depending on the crust, or upper surface of the marsh, must be liable to some degree of uncertainty, and great care in the examination is necessary, in order to determine whether there is sufficient substance to support the railway. The marsh may be too soft; if not on the surface, at some point below, which is sometimes more filled with water, and softer than the surface, and by yielding under pressure, causes the crust to break in. The surface of a marsh sometimes, by the growth for ages of aquatic plants, acquires a considerable degree of firmness, while the sub-stratum is very soft, and so intermixed with water, that it is easily displaced by pressure on the surface. This may generally be ascertained by careful soundings, noting the pressure required to force the iron sounding-rod at different depths as it passes through the marsh.

By a process essentially similar to that first described (by fascines), the Liverpool and Manchester Railway, in England, was carried about four miles across the celebrated "Chat Moss," which has been represented as a very soft, spongy marsh, but not as soft as some we have had to contend with in this country. It was susceptible of improvement by ditching; and after the drainage it received from the railway works, a large portion was reduced to cultivation. In 1850, I saw hun dreds of acres of nearly matured wheat, barley and oats, of luxuriant growth, waving on each side of the railway. This proved the marsh to have had a good degree of substance, which, being consolidated by drainage, has, with the aid of fascines, supported a heavy railway traffic for over thirty years. George Stephenson, Esq., the engineer of this work, showed great boldness and self-reliance in the undertaking, and no doubt had many anxieties during the progress as to whether or not it would answer the purpose, and is eminently entitled to the credit of success. But it would not be safe to conclude that all marshes could be successfully treated by the same process; at the same time, it should lead the engineer that meets this kind of obstacle to the most thorough examination to ascertain the probability of his being able to cross a marsh by works that may be supported on its surface, as this is by far the least expensive.

If the country surrounding a marsh is too high above it to permit bringing the grade level near its surface, and a heavy embankment is required to carry it across, there will be little probability that the crust will sustain the weight. If the soundings show the marsh to be shallow, it may be filled in the usual way, by which the material will force the soft mud, until it finds support on the solid substance below. But if there be found great depth of vegetable matter, it will be important to conduct the filling so as to make the marsh material available as far as practicable, and thereby reduce the quantity. The soft, vegetable matter and fine mud, fully saturated with water, will yield laterally before the bank filling, and the surface of the marsh on each side will rise from the pressure of the solid earth filling, as it passes toward the bottom of the marsh. If the marsh material can be kept under the filling, instead of being pushed aside, it will be compressed until it is capable of sustaining the weight of filling; and though it will be much reduced in bulk, it will be an important saving in expense. A narrow bank, dumped on a soft marsh, acts like a wedge, on the semi-aqueous material, to force it laterally, and displaces horizontally what it is desirable to compress. To guard against this, make the base of the filling wide in the commencement, and fill it evenly on the outer parts, keeping the centre of filling below the sides as much as convenient. According to the texture of the marsh, this floor of filling should be extended beyond the width required for the filling, ten to twenty feet, and carried up as far as may appear necessary to prevent the marsh rising outside the bank, but not to such weight as will break the crust of the marsh, unless the latter is very soft. This extension will tend to hold the marsh material in place, and cause it to settle under the filling, instead of moving it laterally, and thus secure the compression of the marsh material under the bank filling. In the process, care will be required to carry the outsides of filling, with the extension filling, forward of the pressure on the central part, so that the soft earth will move from the sides toward the centre, and finally the central filling will compress it vertically instead of laterally. The compression may reduce the bulk one half or more, but still it will be an important saving in the filling that would otherwise be required in the base, and the bank will come to a rest much sooner than if it is so formed as to force the marsh material outward. The plan of spreading the bank, if the marsh is not very soft, and thus guarding against lateral movement, will reduce the general sinking of the bank, and in some cases will prevent the breaking of the crust, and save much expense. The engineer cannot be too cautious in examining by thorough soundings of the marshes in his way. Sometimes boring may be advantageous; but in general a good sounding-rod will enable him to judge of its character. In the haste that often prevails in commencing the construction of railways, and from the fact that marshes only occur occasionally, this preliminary examination is often very imperfect, and the evil is not understood until the work has made considerable progress; the crust often sustaining a considerable height of bank, and then yielding suddenly, the bank wholly disappears, leaving only water to mark its track. At this stage of the work, if the marsh be deep and of considerable length, it may, and probably will be, the best plan to proceed at once with extension banks on each side, so as to arrest the lateral movement as much as possible. If they be short and not very deep, it will probably be best to continue the filling as begun, which will result in building a bank from near the foundation of the marsh.

The width of embankment has often received too little attention, hardly forming in some cases sufficient support, when new, for the cross sleepers, and the deterioration of a fresh-made and insufficient bank, by settling and the wash of rains, will require constant repairs to maintain the track in operation. After allowance for settling, the top width of bank should extend not less than three feet beyond the end of cross sleepers on each side, to give firm support to the track. Care should be taken to give an inclination on the top, from the centre each way, to carry the falling water quickly to the sides, and prevent it as much as possible from sinking into the road-bed. With nine feet cross sleepers this gives a bank of 15 feet width, or about five feet outside of the outer rail on common gauge, and may be considered sufficient for banks raised only a few feet above the natural surface. This width, however, is not sufficient to afford adequate protection against a car or engine that has left the track from running off the bank; nor is this important on low banks, not of sufficient height to upset the engine or coach. The occurrence of railway vehicles leaving the track is comparatively rare; still it does occur, and at times on an embankment, the car or engine passes over the margin and upsets on its slope, doing much damage to persons and property. According to the height of embankment and the exposure in this respect, attention should be given to secure reasonable safety. On the Hudson River Railway between New York and Poughkeepsie, the width established in the construction for the river side of the track was ten feet outside the outer rail. The object was to prevent a car, on leaving the track, from falling into the river. Unless the train is propelled with unusual force, this is believed to be sufficient to prevent it from passing off the bank; and after ten years operation on that line, I have not heard of an instance occurring over the ten feet margin. It is obvious that this question depends on the speed of trains, and the curvature and condition of the railway and machinery. Near all bridges of considerable height, a margin of ten feet is as little as can be regarded reasonably safe; and to give proper security in this respect, all banks of sufficient height to upset a car in passing off the track, should have a margin of seven feet, and if of great height or otherwise exposed to great hazard in this respect, they should have a margin of ten feet. It is well known that little regard is paid to this feature in our railways; and for the most part they have experienced only occasional damage. Still serious accidents have occurred for the want of adequate marginal width, demanding the exercise of greater care in guarding by all reasonable precaution against their recurrence. The expense is the objection to this view of marginal widths, and in very high banks, will be considerable; especially if the material for filling is not convenient; and want of funds will be likely (as it has done) to induce the narrow and less safe margin; but the instances that often occur, where material may be moved at moderate cost for the safe and proper width, should be improved to give as much safety as may be practicable in this respect. On railways of moderate traffic, where high speed is not required, a margin of five feet will generally do very well, and as traffic and speed increase, it should be enlarged in a corresponding degree, for all high and hazardous banks.

In carrying a railway through a forest of heavy timber, all trees that by falling would reach the track, should be felled and cleared off. The best method is to clear all the timber within 100 feet of the rail; this distance will ordinarily protect the track from the danger of falling trees, and by opening the railway to the sun, be of great benefit to the track.

CHAPTER VI.

BRIDGES AND CULVERTS.

UNDER this head may be classed such works as are necessary to provide for the passage of water, that, by natural streams or artificial channels, flows across the railway. The term culvert is used for works on small streams, and bridges for large ones. No exact line seems to have been settled, to determine the dimensions that constitutes the extent of opening for a culvert, and where a bridge begins; but the true culvert is usually applied to arched openings not exceeding 20 feet, and covered by an embankment. The same style of work, however, is sometimes carried to openings of 30 feet, when the parapets are merely retaining walls for the embankment over the arch, and may with propriety be styled culverts.

There is no material for culverts equal to sound, durable stone. In a climate of moderate temperature, good hard-burnt brick will answer very well; but where severe frosts act on the structure, stone is much better. So far, therefore, as the formation of the country will permit (and a location that makes suitable provision for drainage will usually permit it), there should be provided room for arched culverts. Where only a small opening of one and a half to two feet is required, or even two and a half feet, and sound blocks of durable stone can be had for covering, they may be used advantageously instead of arches. If such covering stone cannot be had, cast iron may be used, and is considered second only to stone; but if durable stone, and a good article of hydraulic cement can be had, it is best to adopt stone-work, either on the square form or arches, as the material may require. With stone of fair, natural beds, laid up in the style of rough hammered work, in hydraulic cement, culverts may be made that will stand indefinitely. In this way they will be of moderate cost for workmanship. If the stone were squared and dressed to courses, the work would be more stable; but this expense is not often necessary in railway culverts, if a bank of several feet in depth of earth lies on the arch, and between it and the superstructure of the railway. Care should be taken to place the floor of the culvert low enough to give full freedom to the drainage of the water course, and if the foundation is of wood, so low that water will always cover it. Wood will not answer for a foun dation that is dry a portion of the year. If liable to this contingency, the foundation should be of stone, which may be made by forming a pavement across between the abutment walls, as a floor for the water course, and should be in the form of an inverted arch, which is more firm in itself, and greatly improves the supporting foundation of the abutments. If the situation does not admit of timber foundations, it will not answer to use wood for sheet-piling, and stone-masonry must be substituted. The foundations for culverts should be perfectly solid, or so as to give adequate support, and if to support a bank of great height, formed in the usual mode of slope sides, care should be taken to give strength of foundation in proportion to the pressure it is to bear, so that if any settlement take place, it may be uniform. It most generally happens, that a good and sufficient foundation may be formed on the natural ground ; the light and yielding surface being removed with other excavations required to reach the necessary depth for drainage ; but if such is not found, bearing piles should be driven to such a depth as may be required to secure stability. The foundations of bridges and culverts are exposed to injury from floods, and should be thoroughly protected in this respect. In all cases where the water is liable to get under the foundation, sheet-piling should be put down; this will generally be necessary at the head of the culvert, and in some cases a course will be required at about one-fourth to one-third of the distance below the head, which should be carried up close to the rear of abutment walls, or what is better, a wing buttress of masonry in connection with and above this sheet-piling and projecting about three feet beyond the rear face of abutment walls, and carried over the crown of the arch, rising about two feet above it, so as to effectually cut off any water that might find a passage along the rear of the masonry, and between it and the earth filling, which is liable to happen when floods raise the water at the head so as to give considerable pressure. Sheet-piling should be very thoroughly put down at the foot of the culvert, to prevent undermining by the action of the water as it leaves the culvert; in many cases it will be best to substitute a wall at this end for the sheet-piling, sinking the foundation as low as circumstances may require, to place it below the power of disturbance from the discharging water. The filling of earth next the culvert should be done with a view to make a water-tight connection, and not with sticks, stumps, stones, and earth indiscriminately, as is too often done. The failure of culverts often arises from currents of water getting under their foundation, finding a passage behind the walls, or from undermining at the lower end; hence will be seen the importance of carefully attending to these details, the neglect of which will very probably be followed by the loss of the culverts, and much additional damage. It is always well to place foundations low, especially if they do not rest on bearing-piles or rock; by this course, all foundations in permanent streams, however small, may be covered at all times with water, thereby protecting any timber in the work from decay.

It not unfrequently happens, that suitable stone for culverts cannot be had at reasonable expense, at the time the railway is in construction, and timber must be used until transportation by the railway afford the means to procure it. In such cases, and if the stone can be procured after the railway is open, the timber, or small bridge, that may be temporarily adopted, should be made with sufficient opening to allow the culvert to be put in, and the earth filled over it, without disturbing the traffic of the railway, which may easily be done. If the foundation be such as to require bearing-piles, it will be best to put them in before the timber bridge is built, as it will be more convenient to drive them before than after the railway is in operation. As soon after the railway is put in operation as circumstances will permit, the work of collecting materials and putting in the stone culverts should be proceeded with; commencing with those, that from any cause may be less safe, and carrying forward the work as fast as economy will permit so as at least to secure completion before any portion of the timber may fail.

To construct culverts of the character here contemplated, the mason-work must be well done, the stone well shaped and laid in good bond, and the joints full bedded in hydraulic mortar. If the engineer is careless on this point, the work will probably be poor. A mason is very apt to devote his attention mainly to the outside face of his work, leaving the interior poorly bonded, and the mortar joints not half filled, and as far as filled, done with untempered mortar of inferior materials. It is not difficult to find samples of pretended mason-work that offer no good reason for substituting it for timber-work. In such cases the engineer is at fault, and not the principle of stone culverts. There is very small comparative difference in the expense of good and poor masonry, and the engineer who allows the latter, cannot be justified on professional or any other grounds.

Care should be taken in filling a bank over a culvert, to load it equally on each side; where it is done by a head bank, as is usual when the material is brought on by a railway, there is great danger of pushing the culvert out of form, if counterweight be not brought to support it.

There have been many instances of failure in railway culverts, and this has led some persons to suppose they are unsafe, and that it is better to adopt timber bridges. It can hardly be said that this substitute more rarely fails than culverts of stone. That there has been a good deal of failure in both is beyond question. But I have rarely known a well built culvert, with sufficient water-way, to fail, though instances have occurred where the water, during floods, has risen at the head from ten to fifteen feet high. The real difficulty is, that the culverts of late years have been to a great extent poorly built, imperfectly founded and protected, and often of inadequate dimensions for water-way. A stone culvert, like any work of human hands, may fail without fault, but well-built stone culverts are works of permanent and reliable character—more safe for the passage of trains, and almost free from expense of repairs; but if the railway company have not the skill necessary in their employment to construct them properly, it will be better for them to procure it, and not trust works of this importance to incompetent men.

In this country bridges are mostly constructed of wood; in England, brick is very extensively used, and in that mild climate may answer a good purpose. So long as bricks continue durable, they give similar stability to stone; but in the northern States of our country they cannot be recommended. In any situation where good building stone may be had, it admits a cheaper structure than brick; it has fewer joints and is less exposed to fracture. Timber is generally used for its cheapness, and the facility it affords for rapid construction. Such bridges are sometimes constructed with stone abutments and piers, but often with frame-work of the same materials. It cannot be denied that in many cases in this country, considerations of expense absolutely control this question, and leave the engineer no choice. It is nevertheless a matter of importance, both in regard to the permanent economy of a railway, and of its safety in use. Timber has only a moderate degree of durability, when exposed to the vicissitudes of our climate; and efforts for its preservation in bridges in this country have not been attended with any great success. Housing has proved very useful for bridges on common roads, but the danger of destruction from fire by locomotive engines renders this form of security on railways less desirable. If a bridge is covered, the inside that is exposed to fire from the engines should be kept well covered with a wash of lime and salt, as a protection of much value against fire. Covering the top with a decking has proved of doubtful utility, and many are now left to the free action of the elements.

The length of time timber will last in a bridge, especially an open one, is quite uncertain, and there is danger that it will be trusted too long for safety. The first decay will be in the joints of framing, and in the interior of the scantling; this may be to a serious extent, while all exposed to observation appears sound and safe. The traveller on the railway cannot examine the bridge—he must depend on the railway agent, under the proprietary interest in the question; and the agent may be satisfied with the exterior, or from other cause neglect the proper examination, until some train falls through, when it will be sadly certain that it should not have been trusted so long. I have seen the record of four serious disasters the past year from the giving way of bridges, causing the loss of fifteen lives, and injuring forty-seven persons. Others less serious to life have been attended with great loss of property. The present year (1859) will not be less disastrous, judging from its history thus far.

A large portion of the bridging that has been built with timber spans, might be made with stone arches not exceeding forty feet span, and many of these with less span. Where abutments of stone are made, as retaining walls for the bank, and to support the timber span, in such situations as allow sufficient height for arching with stone, it will often be found that stone arches may be erected with small additional expense over the wooden structure, resting on stone abutments and piers. Arches for this purpose may be substantially made with a good quality of building stone and hydraulic cement at a moderate cost. Dressing stone to courses is not generally necessary for the permanence or stability of the work. Arches of unhewn stone, with proper spandrels, have been successfully built of seventy-five feet span, with rise of one-quarter, for the use of common roads. I would not recommend going to this extent without dressing the stone for a railway bridge, but there is not the least difficulty in carrying them to fifty feet with one-quarter to one-third rise according to quality of stone. When materials can be had that are suitable for abutments and piers, they can generally be found sufficient for the sheeting of an arch. In making arches of unhewn stone, a common error should be avoided—namely, dressing the ring or heading courses; which not settling with the rest of the arch, is very likely, in arches of considerable size, to split off and separate from the main body. This is mostly done for appearance, and is sadly at the expense of stability. The heading should be of the same workmanship as the body of the arch, giving it no more care than would be given to the face of an undressed wall.

There are comparatively few situations where arches of more than 50 feet are required, to provide sufficient opening for the water-way. If one arch is not enough, two or more may be provided. Small arches are less expensive for the span they provide than large ones; and hence it is often better to make two arches of equal aggregate span than one, to provide a given opening; and they require less height. The size of the opening for water-way will depend much on the exposure to obstruction from drift-wood or ice; and it will require careful observation to settle this on a safe basis. It is not often serious on small streams; but for large streams, exposed to heavy floating ice and drift-wood, the safety of the structure will depend materially on the proper disposition of this question. The opening that may be necessary to give *free passage to the water*, is the first question to be considered in planning a bridge or a culvert. It should be sufficient to pass any floating substance likely to be brought by the current to the bridge. On this point it is best to err on the safe side, in order to be prepared for that great flood, more weighty than any previously known by the "oldest inhabitant." A water-way barely adequate to pass a flood will severely try the foundation; and it should be kept in view, that occasionally—perhaps once in a quarter, or half a century—streams of water are swollen much beyond ordinary floods, and sweep off what had been regarded as well tried and safe. The water-way may require two or more spans to provide sufficient flow, but each should be wide enough to allow any ice or drift-wood that the stream may bring down to pass through freely, so as not to dam up and obstruct the flow of water. In regard to ice, if the stream directly above the bridge is rapid, the ice will be broken into small pieces, and pass off with little hazard of forming a dam at the bridge; or if it be very crooked and sluggish, the ice will be held in its original position, until it becomes too weak to cause much obstruction in passing the bridge. There are comparatively few streams that would not be safely secured against ice or drift-wood forming a dam, by spans of 50 or 60 feet; and if more is required to give sufficient water-way, the number of spans can be increased accordingly. If larger openings than 60 feet are required, and there be sufficient height, stone bridges may be advantageously erected, but will require more expensive workmanship and materials.

The remarks in relation to the masonry of culverts are applicable to that required for bridges. For bridges of unhewn stone, the stone should be larger in proportion to the magnitude of the walls and the pressure they may be required to sustain. The judgment and experience of the engineer must decide, as to the pressure from the height or span of the work required, how far he may with safety adopt rough or unhewn stone-work. So far as this can be done, it will be much more economical than hewn stone; and, as before remarked, it will be found in general that by far the greatest proportion of bridges may be constructed of unhewn stone, without sacrifice of any material stability. It sometimes happens that the stone quarries that must be resorted to will furnish stone in such shape, that it is little more expensive to make regular courses roughly hewn than rubble-work, and in such cases it will be advisable to form courses of uniform thickness of stone for the more important features of the work.

Stone arch bridges require more height, or space below the grade level, than timber; and it is to be considered whether or not this can be obtained without too much sacrifice in the cost of grading. It sometimes occurs that the formation of the approaching country, on which the line of railway is laid, is low in comparison with the bed of the valley or stream over which the bridge is required to be made, and the most favorable grade for the approach does not give space for the stream in time of floods under the bridge; and the expense of raising the grade will sometimes so control this question, as to lead to the adoption of timber or iron for the want of room for arching. The grade should be high enough to be beyond the reach of all floods in the stream, and a few feet additional is all that is necessary for arches of small span; and when this can be obtained at moderate additional cost for the filling, arches of stone should be provided for. In examining the circumstances of grade with a view of ascertaining what room may be had for arching the bridge or culvert, it will be kept in mind, that raising the grade will improve the facilities for drainage, a matter never to be lost sight of in the construction of a railway, and will always justify some expense to improve it, even when otherwise fair.

It must be conceded, however, that there are situations, where a railway passes over a flat country, with shallow valleys for its streams, where timber or iron must be used for the spaces or spans between the abutments and piers, for the want of room to put in arches of stone; but the cases are not comparatively numerous in which this may not be avoided by a judicious management of the grade lines. Cheap construction in the outset will call for low grades in a flat country, but the true and permanent interest of the work requires them to be higher than they generally are, both in relation to passing streams with safety, and to give effective drainage. Low grades, and consequent imperfect drainage, have been a very prevalent error on the railway constructions of this country. The engineer should keep this in view, and avoid the error.

If stone for bridges is not to be had on the line of railway, then the remarks in relation to culverts in this respect will be applicable to bridges, namely, the foundations prepared, and the temporary timber structure so constructed, that when materials can be brought on, stone-work may be erected without disturbing the operations of the railway.

When there has not been space enough for arches of stone, as before observed, timber has generally been substituted; but this is not indispensable, as beams of wrought iron may be constructed for supports between and resting on the stone abutments and piers; and if the span is not to exceed 50 or 60 feet, will not be very expensive. Such beams should be strong enough for the service without counter-supports, in the manner of a simple beam. Make two beams of boiler-plate iron, one for each side of the track, and by rods from them support the cross-beams on which the rails of the track lie. The cross-beam may be of iron, and no timber used in the structure, except for the stringer on which the rails are laid. Such a structure will be free from the objection that arises from the expansion and contraction of iron, as the action of the temperature does not change its character for support, and may easily be provided for. For a short span, not exceeding 40, or perhaps 50 feet, a single sheet of iron, with flange pieces at top and bottom to give lateral stiffness, and secured with angle iron, may be made. If the span is greater than 40 or 50 feet, the beam should be cellular, having two upright sides, and connected at the top and bottom with plates secured with angle iron. The depth of beam should be about one-tenth of the span, and the breadth about one-sixth of the height. This would be a tubular beam, to which the cross-beams would be attached as above described. It has the same principle as the tubular bridge; it is equally firm for support, but the span must not be carried too far, or beyond its means of lateral stiffness. The objection to this plan is, that the solid sheets in the side plates contain an unnecessary amount of material. With sufficient strength in the upper and lower chord, all that is wanted between them is what will be sufficient to hold them firmly in position, and this may be secured by a system of bracing requiring much less material than the side sheets. It is however a question, if the extra workmanship in the bracing will not require greater expense than can be saved in the material of the tube, as is claimed by Mr. Robert Stephenson. Mr. S. must be regarded as high authority on this point, and should not be disregarded without careful experiments that might demonstrate the economy of a modification. There have been some experiments on the plan of connecting the upper and lower chords by bracing, on the New York Central Railway; which, so far as I have been able to learn, has satisfactorily stood the service of trains for some time; but I do not know what difference there has been in the expense as compared with full sheeted sides.

Railways must occasionally be carried across large rivers, that demand works of a different character from what has been contemplated in the preceding remarks. In some cases, circumstances render it necessary to make a greater span than could well be made by stone arching. Timber has been mostly used for such cases in this country. In England, the tubular wrought-iron bridge has been used for the greatest spans; and the Grand Trunk company adopt this for their bridge at Montreal, over the St. Lawrence River, and for many smaller bridges on their line. The latter plan is, no doubt, substantial and good, but has the objection of being expensive. The wire suspension bridge has been adopted to a limited extent in this country for railway purposes; as at Niagara Falls. An iron truss bridge, invented and patented by Mr. Fink, an engineer of the Baltimore and Ohio Railway, has been several years in use on that line, and is spoken of favorably; I have seen a copy of the plan and specifications, and think it the best for an iron truss that I have seen. It has a very satisfactory arrangement to provide for the unequal expansion and contraction of large and small pieces of iron, whereby the different stages in time of change in large and small bars, and of cast and wrought iron, are equalized by an adjusting motion.

The suspension plan cannot be considered as favorable for railway use as the rigid plan, and the former will not probably be adopted, except for large openings in situations that do not admit of division by piers, as the one at Niagara Falls. Where there are reasonable facilities to found piers, and circumstances do not admit of stone arches, iron in some form may be advantageously adopted. In considering any plan for an iron truss it is indispensable to see that provision is made for the unequal expansion and contraction by changes of temperature: wrought iron and cast iron do not work equally, and the large and small bars will require different spaces of time to reach their limit of change by the same degree of temperature. This, as before remarked, seems to be well provided for in Fink's plan, and which is certainly very ingenious. But the open tube has the advantage of not needing any such provision, the metal being of uniform kind and thickness, its variation in length is all that requires attention, and for this it simply rests on rollers; and being simple in construction, will be likely to obtain preference when iron bridges of large span are wanted. Iron used in bridges should be carefully and well covered with paint to protect it from oxidation; and this should be watched, and renewed as it wears off by the action of the weather.

Pile bridges are sometimes used; but they cannot be recommended, and should only be adopted as a temporary expedient, to hasten the opening of the railway, after which they should be superseded by more substantial and durable structures.

The observing railway traveller cannot fail to see, that, as a general thing, our railway bridges are far short of the stability necessary to their safety and the true economy of such works. The rapid demand for railway accommodation, and the limited means at command for construction, have doubtless led to the erection of temporary and perishable structures. But these should give place to those of a more substantial and durable character as fast as circumstances permit. It is not to be supposed that timber bridges can, in all cases, in the construction of a railway, be dispensed with. Circumstances of a controlling nature will often render them expedient; and all the engineer can do, is to so plan and prepare his work, that this perishable material, unsafe from fire and ordinary decay, may at the earliest period give place to durable structures, that will be safe for travellers, and best promote the interest of the proprietors.

At this time there are numerous timber bridges on our railways, and many of them have stood as long as they should be trusted; and it is quite time they were replaced by stone or iron—the most suitable, if not the only suitable materials for such purposes. These wooden bridges are dreaded by all passengers, who are aware of the uncertainty of their strength, and the danger of their failure under an exterior appearance of soundness, when, in fact, they are rotten at heart. If those whose business it is to know, are often mistaken as to their condition, how shall passengers understand if, or not, they are safe? But they are sometimes used, when those in charge know that they are not only hazarding the property of the company, but the lives of the passengers. It is hoped and believed the latter fault is not often practised; but whether from error in knowledge, or lack of fidelity, the hazard is the same. It will properly be said, there is no excuse for this; the man in charge should be competent and faithful. Now, men are men, and the true way is to construct works of such great importance, as far as practicable, in a manner that will need the least possible watching.

CHAPTER VII.

ROAD CROSSINGS.

THE more prevalent practice in this country is to make the highway crossings on a level with the rail track—bridges being the exception and not the rule; while in England, a level crossing is rarely seen; and where they do occur, gates made of strong bars of scantling are put up, and so arranged as to close the highway when the track is open, and *vice versâ*, so that no passing can occur on both at the same time. The highway is usually kept open; and when a train approaches, the man that tends the crossing shuts the highway and opens the track until the train has passed. I do not recollect seeing this gate arrangement at crossings in this country, except at some of the railway crossings in the streets of Boston.

There are many situations in which a bridge to pass a highway over a railway track would be very expensive, and this is a strong reason for the level crossing; and in some situations near stations there would be inconvenience in the approaches to the railway. That highways should pass over or under a railway, where circumstances permit it, no one will question. The level method of crossing has caused much personal injury, the loss of many lives, and many persons maimed for life, and much damage to property. It is therefore incumbent on the engineer to consider well every facility the location presents, of guarding against the necessity of a level crossing, especially where there may be obstruction, preventing a good range of sight. This subject is of sufficient importance to justify some modification of grade or line, if by that means a highway may be passed by a bridge. There are situations that favor bridging over or under; in some cases, the route of the highway may be changed, so as to permit bridging without serious expense. To raise an embankment to the elevation required to pass over a railway, sometimes interferes with other avenues or objects to such a degree, that it cannot well be done; but if there be nothing of this kind, and the expense of raising the bank for landing the highway over the railway be the only objection, the case should be very unfavorable to warrant a level crossing. The objection to merely an elevation of the highway as injuring the public accommodation for travel, is by no means worthy of being placed against the benefits the travellers on the public road will feel, in the safety from collision that will be secured by the bridge. If a level crossing must be adopted, the guard gate should also be adopted to prevent collisions; especially if there be not a good view by which parties on the railway or the highway may readily see each other.

The same reasoning, in a modified sense, holds in relation to farm-road crossings. Serious accidents have occurred at these when made on the level of the rail. As they are numerous, and less important than highways, they are more generally made level; but, so far as reasonably practicable, it is better to adopt bridges for these. I am well aware this subject has received very little of the consideration due to its importance, and that the expense of the methods recommended will be regarded a paramount objection, except in quite easy cases. If all the damage the level crossings have cost could be collected together, I am sure it would show, that as a mere question of economy the practice should be less common. In addition to this, there are the uncertainty and consequent anxiety that must be experienced by the managers and on the part of travellers, especially if trains are run at high speed.

To the guard gate it will be objected, that it will

involve the expense of an attendant keeper, more or less of the time. Doubtless it will involve this current expense; but it will save life, and prevent damage to the railway proprietors, that will eventually be equivalent to the extra cost. The expense of a guard gate-keeper is a legitimate consideration to be weighed in favor of a bridge.

In this, as in all matters relating to railway management, there should be a careful consideration of the means that may relieve the chances of accident. It is not sufficient that it is some man's duty to watch and prevent apprehended danger; for men supposed to be careful, and who for some time may be so, yet from the daily routine of a specific duty, are liable to become listless, and however we may complain of their neglect of duty at the important time, the mischief is done. No doubt, some men are more negligent than others, and the most rigid scrutiny should be exercised in their employment and supervision; but men will be men, and as infallibility is not to be expected, the works of a railway should be arranged, as far as reasonably practicable, so as to avoid the necessity of special watching. There will be ample field for this duty, after the works are arranged in the most favorable manner.

CHAPTER VIII.

BALLASTING TRACK.

There are two objects to be secured by ballast; first, a material that by its open, porous character, will allow falling water to pass off freely, and prevent heaving by frost; and, second, by its hardness and firmness in all states of weather, to sustain the action of trains, so as to prevent, as much as possible, the derangement of the track. If these conditions are secured, all others will be essentially provided for, so far as depends on ballast. In districts abounding with clean gravel and sharp sandy soil, a good material for ballast is easily procured, and may be put on at the time of grading the railway. But it more generally happens, that suitable material for ballast cannot be had in sufficiently close proximity to allow profitable hauling by teams, and it will be necessary to haul it by trains on the track. This may be done by laying the track on sub-grade, or on the base that is designed for foundation of ballast; it should be done with care, so as not to bend or otherwise injure the rails. Where circumstances permit, it is best to put down the ballast before the track is laid; but if proper care is used, no great objection exists to laying on sub-grade, and is most economical if a track is required to be made for its transport.

The quantity of ballast required will depend much on the material on which it rests, and the depth penetrated by frost; and will range from one to two and a half feet. If there be any difference in the quality of the ballast pit, the most clean, open and hard should be placed in the cuttings; the inferior on the banks, where settlement will be likely to demand an extra quantity, and where the drainage is more perfect than it can be in cuttings. With a well drained road-bed, fifteen to eighteen inches of good ballast will, in most cases, be sufficient.

In many locations, ballasting material is only found at a few points on the line of railway, and requires a long transportation for such a purpose; and sometimes it can only be had by constructing more or less of temporary railway to reach it. This has often led to a postponement of ballasting, and the railway has been opened for use without it. In such cases, the rails are pretty certain to be more or less damaged, if the track be run in this condition for any considerable length of time If in the course of construction it appear probable that such plan must be pursued, the rolling stock necessary to operate the railway before ballasting can be done, should be as light as will reasonably answer for the work to be done. The light stock procured for such purpose need not be lost, or thrown aside after the railway is in condition to bear a heavier class, as there will always be found sufficient business to employ this light machinery advantageously. It is obviously of great importance that, under these circumstances, the drainage should be most thoroughly done. There are many soils, that if well drained would bear the track and sustain a considerable traffic, conducted with light machinery and moderate speed. But to dispense with or defer ballast, can only be tolerated under the pressure of urgent necessity, especially for a liberal traffic.

Clean gravel and sand are the best materials for ballast. Broken stone is very good—it makes the most durable ballast, and has the least tendency to make dust; but is not so favorable for the durability of the rail, or rolling machinery. Formerly it was considered the only suitable ballast, and is still used to a considerable extent where suitable gravel cannot be had. Stone, for such a purpose, should be broken so that any piece will pass a two-inch ring. The quality of stone to be preferred, is a rather friable, hard, silicious kind, rather than a more solid and unyielding quality, as it favors the elasticity desirable in a rail track. It seems a contradiction to require solidity, and at the same time elasticity; but experience has shown there may be too much solidity; traction, or adjustment of track may be improved by it, but the rail and rolling machinery in high speed are too severely acted on, and hence a compromise has to be sought, and experience has settled on gravel and sharp sand as most beneficially securing the respective objects. In England, brick clay has been burned for this purpose, and is said to answer very well; and where suitable clay is found, and coal or wood can be had at moderate cost, it may be advantageously used in this country, where the usual material cannot be had.

The width of ballasting should be such as to give a good support to the cross sleepers, and prevent them from moving laterally, and if it is intended to give the track a good margin for safety on embankments, the ballast should be full width. In cuttings, it is not necessary to extend more than one and a half to two feet beyond the ends of the cross sleepers; no more width being wanted than necessary to protect the sleepers from lateral movement, and except in sharp curves they are rarely known to move, if the ballast on the top extends so as to cover the ends of the sleepers, and slopes off to one and a half feet beyond.

Ballast is necessary on all soils that do not of themselves constitute a good ballast: it is necessary in rock, to give the degree of elasticity found to be necessary on a rail track. On a soft base it is a good plan to lay a course of broken stone, and complete the ballast with gravel. The injurious effect of too great solidity in the support of the rail has been alluded to, and this is strongly manifest where the road-bed is made solid by frost. The effect of frost is in a measure counteracted by good ballast, as there is more elasticity in a free, porous ballast than in a retentive soil when frozen.

CHAPTER IX.

CROSS SLEEPERS.

It was formerly regarded necessary to the proper stability of a railway, that it should be laid on stone blocks resting on a bed of broken stone. This plan gave great firmness to the structure, and easy traction to the machinery; but as speed came to be increased, it was found to operate severely on the rails and rolling stock, and has given place to timber sleepers. Some experiments have been made in England on cast iron as a substitute, but as yet very little has been done toward displacing timber. The elasticity of timber increases the tractile force required to haul a train; but this elasticity is too important to be disregarded, if high speed is to be maintained; this circumstance, together with the facility with which timber can be used, gives it a great advantage for this purpose, and notwithstanding the objection from its liability to early decay, it will probably be used for a long time to come.

As rails are laid in this country, resting their

web base directly on the sleeper, without the intervention of a chair, except at the joints, it is desirable the wood should be hard, to prevent the rail from sinking into the sleeper. What are termed soft woods, will not long sustain this service; and what is wanted in a sleeper is, durability with sufficient density to bear the action of the rail. The kind of sleeper most generally preferred in this country is white oak. Chestnut and Chestnut oak are good timbers. Locust and red cedar, though not as hard as oak, from their great durability and fair density, make good sleepers, and if the weight of machinery is not too great, would be the most valuable kinds; but these timbers do not abound, and can rarely be obtained at suitable cost. Red elm makes a good sleeper, hard and durable, scarcely inferior to white oak. Black cherry and black walnut answer very well. Other kinds have been used. The engineer will most generally be compelled to use, in original construction, the timber that can be obtained not very remote from the line of railway, and should take the best that may be had.

In England, and in Europe generally, timber is much more expensive than in this country, which has led to various methods for preserving it from decay; and from the best information I have obtained, they have succeeded in giving the sleepers about double their natural durability. This is a very important point for them, where the renewal is attended with heavy expense. The cost of the preserving process is variously stated for the different methods, from ten to twenty-five cents per sleeper, and shows an important saving secured by the increased durability. The cost of sleepers in England may be taken at not less than one dollar each, and therefore the object of preservation is much greater than in this country. Timber, however, is becoming dearer with us, and attention will soon be, if not already in some parts, demanded for its preservation. On the methods that have been adopted to increase the durability of timber, much information is given in the work before quoted, "Colburn and Holly on European Railways."

Sleepers should be well seasoned before they are laid down, in order to obtain the greatest natural durability, and to hold the spike well that secures the rail. Green white oak is very unfavorable; it will, in a short time, so corrode the spike that it may easily be drawn. If care is taken, as the sleepers are got out, to pile them properly, a good degree of seasoning may generally be secured before they are required for the track; this should not be neglected, though I think it often is.

The most usual method of getting out sleepers is to hew two sides parallel and leave the others round, or in their natural state; this makes the best sleepers used in this country. Sometimes the bark is removed from the unhewed parts, which improves their durability. They are sometimes sawed or split out of large timber, but these are not so good or convenient for adjustment of track as the flatted sides. The English mode of sawing is better than ours; the sleeper is first sawed in a square for two, and then sawed diagonally, so as to produce two triangular sleepers; and being laid with the large angle down, they are easily adjusted. The width of the flatted sleeper should be considered the heart, or exclusive of the sap of the wood, as the latter will soon decay and should not be relied on for the support of the rail. The timber being hewed only on two sides, it should be reasonably straight, so as to give steady and firm support to the rails.

For the ordinary gauge of track, sleepers have usually been from seven to eight, and sometimes nine feet long. The latter seems most suitable, as giving an equal support each side of the rail; but I have known very good tracks well and easily maintained with sleepers eight feet long. The usual length of rail is eighteen feet; a joint sleeper ten inches wide and ten feet long, with seven intermediate sleepers eight feet long, with a minimum width of seven inches, and six and a half inches thick, makes a good support. A minimum of seven inches will give an average of eight inches in width, and should be laid so as to give, as near as may be, a uniform support to the rails. This number and size of sleeper makes them as close as is convenient for repairs. The Goshen line of the Michigan Southern and Northern Indiana Railway was laid on this plan of sleepers, and under a high rate of speed, proved very satisfactory, and may be considered a good basis for a first-class railway. For branch lines, and railways designed for a moderate traffic and moderate speed, the cross sleepers may be proportionably reduced in size.

Sleepers are in some cases laid on longitudinal sills, but, except on soft ground, and where the railway is to be put in operation without ballast, they are worse than useless. In a few cases, where it is important to spread the bearing of the track over a marshy or other spongy foundation, they may be necessary. At this time they are rarely adopted.

To some extent, timber stringers have been laid on the sleepers, and the rail laid on the stringer; on this plan the sleepers are laid further apart. This is not so convenient a method for adjustment, as that of laying the rail in the usual way, directly on the sleeper; it increases the perishable part of the superstructure, and is now rarely seen in this country. I do not think it has advantages to counterbalance the objections, except where a very light rail is adopted for a small specific traffic; and even in this case, if timber is not very cheap, it is believed a light T rail on cross sleepers is the better plan; and for a railway of good general traffic, there does not appear room to doubt the superiority of the usual practice.

CHAPTER X.

CHAIRS AND SPIKES.

In this country, chairs are used only at the joints or ends of rail. In England, they use a chair at every sleeper. This requires for the American rail a broad, flat bottom, or web-footed base, to give the requisite bearing surface on the sleeper, to prevent the rail cutting into it; whereas the English rail is the same, or nearly the same form on the top and bottom, and is wholly supported by the chair. A chair on each sleeper gives greater facility in furnishing a broad bearing on the sleeper, and allows softer wood to be used; for the chair may be broad, and the action on the rail will not so soon cut into the sleeper. Rails are sometimes laid without chairs at the joints, depending wholly on the spike to keep them in place—a plan that cannot be recommended. The chair at the joint is necessary to hold the ends of the rails firmly in place, and cannot with prudence be dispensed with, unless the fishing joint be used.

A great variety of patterns for chairs, both of

cast and wrought iron, has been used, some with wooden keys and some without. Of late, the most usual chair has been some form of wrought iron without keys. A wood key saves the expense of close fitting. It is hardly possible to obtain so close work, between the manufacturer of rails and of chairs, that there will not, in many cases, be more or less work required to make a good joint, or one that will be reasonably close; and no care will make the track so good and free from rattling sound as may be secured by a wood key. Wrought iron chairs are the most safe, but, in order to economize their weight, so as to bring them near the cost of cast iron, they have often been made too light to be as good as they may be made without wood keys, and I have never seen one with a key. To make wrought iron chairs to receive a key, would make them still more expensive. It is not so easy to give the proper form of chair to receive a key in wrought, as in cast iron. Though a key would add to the safety, and take away the rattling sound of a track laid without, still there can be no doubt that a wrought iron chair of the most approved pattern, well laid, holds the rails very firmly together, and makes a good track. What is now wanted is a chair in some form that will take the bearing of the wheel, and prevent its dropping at the point of the rail when it is opened by contraction, or drawn out of place, and so prevent a blow that batters down the ends of the rails. This would considerably increase the expense of the chair, but would serve a very important purpose in preserving the ends of the rails. This feature will be further considered in the discussion of rails.

On the subject of spikes, little need be said. The form of rail generally adopted in this country requires the brad, or hook-headed spike. The form is well understood, and they will require to be made of the best quality of iron. They are usually from half an inch to five-eighths of an inch square, and four and a half to five inches long, clear of the head; which is quite sufficient for hard wood sleepers. Two spikes are required for each rail at each sleeper, and two extra at the joint will be required to make a good track. Fewer are sometimes used, and may answer for railways of light traffic and moderate speed.

CHAPTER XI.

RAILS.

The first form of rail used in this country, consisted in a timber stringer, on which a flat bar of iron was laid; the iron bar forming the rail track for the wheel, and the timber the supporting material. Though now generally out of use, this kind of railway has done much service in this country, in a very economical way; and paid the proprietors, in several cases, the best dividends, and, in fact, better than when subsequently laid in a more perfect and substantial manner. With suitable machinery, and working at a moderate speed, such rails may do a profitable work; but they cannot be recommended at this day for the service that is now required by general traffic on railways. For short railways, doing a mineral traffic, or for short branches, there may be situations where they could be advantageously used; and, in fact, some are still in use. The wear on the iron is very light; doubtless owing to the elasticity of the timber on which it rests, and the light machinery generally used on them. It has been stated that the timber stringer has been and still is used for a heavy iron rail in some cases. The longitudinal timber, as it is termed, being adopted for its advantage in giving a more uniform support to the rails; but the practice is limited, and rails are now generally made, for railways of general traffic, to rest directly on the cross sleepers.

The difference in the general methods of support between the English and American railways has been noticed. The railways on the continent of Europe have adopted, to some extent, the American plan, but for the most part, I think, they have the English. The former has the advantage of simplicity and ready adjustment; the latter, that of increased means to enlarge the bearing on the sleepers; and it has been urged as a further advantage, that they admit the form of top and bottom of rail to be alike, so that when the top is too much worn for use, they can be reversed and the wear of the two heads obtained from one rail. This is certainly a plausible consideration; but for some reason it does not appear to be much practised, the greater part of English and continental rails on this plan of supports do not have the top and bottom exactly alike; from which it would appear that this is not generally regarded of material importance. The American rail was invented by the late Robert L. Stevens, of New York, and first used on the Camden and Amboy Railway. On its first introduction (about 1830) it was regarded of doubtful utility; it being apprehended the brad-head spike would not secure it against lateral thrust, and if it did, the rail would fail in the waist. These apprehensions have been dissipated by very general experience, and they are at this time almost wholly used in this country. Manufacturers do not like them so well, on account of the difficulty in forming the thin, broad, web base, which is not as easily done as the more compact and rounded form given to a rail that depends on chairs for support. The American rail in dispensing with the intermediate chairs, is less expensive to lay, and has the greatest facility for adjustment; and we may be satisfied with it, as on the whole equal and, I think, superior to any other.

The U rail, as it is called, resembles the letter U inverted; it has been mostly used on the timber stringer, and has some advocates, but has not come into general use.

A great deal of ingenuity has been exercised on the form of rails, both in this country and in England. The plans do not seem to have yet reached a satisfactory result, at least not sufficient to produce uniformity, and modifications continue to be proposed for their improvement. They have gone through many changes, and the best of each kind, so far as relates to the bearing and strength of the rails, may be regarded as pretty fairly adapted to their use. There is, however, one point that approaches slowly to the proper form; namely, the face or track surface on which the wheel rolls. It is remarkable that this has had, to a great extent, and at times almost universally in England and this country, a round form. Whatever advantage this may give to the form and durability of the rail, it is obviously very bad for the wheel; as the tendency is, and the result has long shown it, to cut a groove in the rim or tyre, more rapidly destroying the wheel, and as the groove forms, increases its friction at the periphery, and operating injuriously on the rail. The groove is liable to be neglected until it is unsafe to run the wheel. With a flat top, a groove will eventually be formed, but much less rapidly; and the later patterns are less rounded, and, what is better, some have a flat surface for the wheel. What is wanted is, as much breadth of flat top as can be got, leaving so much to be rounded off, as will give adequate firmness and wearing surface on the edge that receives the action of the flange of the wheel.

The weight of rails that rest directly on the cross sleepers for support, has been increased from

35 pounds per yard, the original pattern on the Liverpool and Manchester Railway, to about 95 pounds. For some time past the tendency has been to reduce from the greater weights, and from 56 to 65 pounds per yard may be regarded as embracing most of the patterns now being made. The manufacture of large patterns is not considered to be as perfect, at least they are more likely to be imperfect than smaller ones; and to this, in a great measure, has been attributed the disappointment in the durability of the largest class of rails. The strength of rail to support the weight that comes upon it, must depend on the frequency of supporting sleepers; and consequently as these are near each other, the rail may be lighter. With such an arrangement of sleepers as has before been proposed, a 60 pound rail may be formed, so as to be sufficient for the service of a first-class railway.

The conditions I would propose for such a rail on the American pattern are: height, three and five-eighths inches; breadth of web, three and seven-eighths inches; thickness of waist, five-eighths of an inch, and a level surface on the top of one and a half inches. These dimensions properly formed to a 60 pound rail may be well manufactured, and will make a first-class railway.

There are lines of railway in operation that have

a light traffic, mostly local; and doubtless there are many other districts, that as the country is improved, will require similar railways to provide for their traffic, and are not sufficient to support a first-class railway; for these lines a rail weighing 45 pounds to the yard may be quite sufficient, more durable, and be adopted with better economy than a heavy rail. A railway of this kind should have all its parts to correspond; engines and cars should be proportionately light. It would not be necessary in such districts to run trains at high speed, and the traffic may be conducted as cheap per ton, or per passenger, as on ordinary first-class railways, and confer great benefits on the routes they traverse. This view of railways will increase in importance as the population increases, in districts that, from the formation of the country, are not and cannot be traversed by leading lines, and can only reach the main lines, or the marts of commerce, by common roads, or by railways that may be supported by a local traffic. As the science of railways comes to be more thoroughly studied, it will be found that a light class may be profitably and usefully conducted, and its benefits extended to many situations that now look hopelessly for such accommodation. Strange as it may seem, it has been the usual practice to put machinery of essentially the same character and weight on railways of light traffic, as is used for those of heavy traffic, even in cases where the track has been made with a rail of lighter weight. This is one of the errors of railway practice, and will doubtless be corrected in time; but as yet the weight on the rail and the speed of trains have not had adequate consideration, and especially on those having a secondary amount of traffic.

The rail, in most cases, fails first at the ends, and a good deal of attention has been directed to find means to protect it in this respect. The effect of temperature to open and close the joint is well known; this is sometimes increased by the want of proper attention in laying down the rails; workmen do not always judge well of the temperature existing at the time; and if allowance is made for too low a temperature, there will be danger the rail may rise up and sometimes be very troublesome to adjust, and dangerous for passing trains; hence, they usually err on the other side, causing greater opening at the joint than is due to temperature. The longer the rail bar, the greater this difficulty. Other causes operate to draw the joints apart, and it is not unusual to see openings of half an inch to one inch, and occasionally I have seen them two inches on railways of large traffic; this is danger-rous and should not happen. To remedy these evils, rails have been made in two pieces longitudinally, and put together so that the joint of one part is supported by the other, by what is termed breaking-joints, and is known as a compound rail. The two bars are riveted together as they are laid in the track, and form, united, the usual pattern of rail. They have been usually divided vertically, so that the joint passes through the head or track for the wheel. Other forms of compound rail have been used; I have seen one that leaves the head in one entire piece. A wheel rolls much more smoothly over a compound rail, and gives a very agreeable motion, free from the jar at the joints. Not only pleasant motion is secured by breaking the joints of the rail, but the effect of the blow caused by the common joint on the foundation of the track and the rolling machinery is avoided. The importance of securing these results is obvious; but the compound rail has not been generally adopted. It does not appear to have the same durability as the common rail, though the joints are much better protected, probably owing to the head or track being divided in two parts, and held together by rivets that are hardly sufficient for the severe action to which they are exposed; and of late they are not as much esteemed as when first laid down, and I think but few of them are used at this time. The full-headed compound rail before noticed, presents the full form of the head, without joint, to the action of the wheel, but does not prevent the joint from opening by the contraction of the rail, but secures it from opening by any other cause, the same as a fishing-joint. I have seen a piece of this rail that had been down on a grade of 80 feet per mile, and four years in use under a very heavy traffic. The bolts or rivets were considerably worn, and many of them loose, but the rail was in fair order for the traffic it had borne; a piece on the same railway, and laid at the same time on a light grade, was in very good order. I do not see that anything is secured by this rail, beyond what is secured by a good fishing-joint.

To guard against the difficulty at the joint, a method before alluded to, and styled a fishing-joint, has, to some extent, been adopted. This consists in riveting two plates of iron, one on each side of the rail, the rivets passing through the rail; the plates are rolled in form to fit the rail at the edges of the plates, and are from 18 to 24 inches long, with two rivets in each end. They are very efficient to hold the rails in line, and are in considerable esteem in England. It is obvious, however, they cannot prevent the action of the temperature in opening and closing the joint, but are effective against other causes of opening. The plates are strong enough to sustain the wheel, and are laid so as to bring the joint between, instead of on the sleepers, as no more support is required at the joint than for other portions of the rail, and consequently joint sleepers are not required. If the outside plate was so formed that its upper surface should come to the level of the rail, and be able to take the wheel at any instance of the opening of the joint, it would greatly, if not effectually, protect the ends of rail from being worn off, and relieve the jar materially at the instant the wheel passed over it.

A method has been adopted to sustain the wheel at the joint, by laying a wooden scantling immediately behind the joint, and having its top level with the rail, the scantling extending from, and resting on the two sleepers each side of the joint sleepers, and firmly secured in place. A hard piece of seasoned white oak is said to bear this service very well, and the plan is much approved by some who have adopted it. It, no doubt, will be much relief to the rail joints, so long as the timber sustains its level with the rail; but it may be doubted if any wood will long bear such service. Such a piece of scantling has been used to form a fishing joint; in which case it is formed to fit the side of the rail, and is bolted to it, but does not rise to the level of the rail; I am not aware that it has been used any great length of time; but it is said to have worked very well. This is merely using wood instead of iron for a fishing plate, and I think it has advantages over iron in the elasticity of the wood, which favors the bolts. It differs from the iron fishing in having the scantling only on one side, whereas the iron fishing has a plate on each side the rail; with an iron plate opposite the scantling, the strength to support the rail would be improved, and without this it is necessary the joint should rest on, and not between the sleepers. As before observed, the fishing of the joint cannot prevent the opening that arises from temperature, but secures it against any further opening, and this is very important. If it be found important that the fishing joint be made sufficient to sustain the weight of the wheel between the sleepers, as is the usual practice in England, the iron plates will most probably be preferred to the timber scantling. I saw at the engineers' office of the New York Central Railway a full-sized model of fishing plates, that I think will succeed well for iron fishing if made a little longer, so as to take two bolts at each end.

On railways of heavy traffic and high speed, it is not probable any method of securing the joints will hold them permanently. The severity of the service will wear and weaken some part, and repairs and renewal will demand watchful attention; at the same time, and for this very reason, it is important to ascertain the best method, and this must result only from carefui experiments on the different modes. The scantling method that has been described now promises favorably, and is worthy a careful trial in comparison with the iron fishing. A chair that will securely hold the rail in place is a very fair method, and will answer a good purpose on roads of moderate traffic, where high speed is not required. The engineer will find this subject one of great interest, and a wholesome tax on his ingenuity and capacity, notwithstanding it may appear to some as having been exhausted. The deterioration of rails, especially at the joints, under the service they have been required to perform, has assumed an importance not heretofore well considered.

CHAPTER XII.

STATION BUILDINGS.

THE convenient arrangement of station buildings is highly important to the transaction of railway business, and demands careful attention from the engineer, especially for the terminal and more important way stations. It is not necessary that they be expensive, but arranged so as to effect the most convenient transaction of business. Local circumstances, as also the nature of the traffic, will have great influence on this question, and more or less modify the extent and main features of the plans. On some lines, the handling of freight is done by individuals, in which case a portion of the expense of station building is, or may be, borne by other parties—a practice of not very general use. For the most part, the railway company do the handling, and if they put it in the hands of others, most generally provide the necessary buildings.

In a *freight station* it is necessary to consider well the plan that will bring the goods to a point, that with the least movement will allow the transit between the railway cars and the vessel, wagon, or cars of other railways, to or from which it is to be transferred; having reference to the labor of loading, unloading, stowing, and taking account of the parcels of goods; amounting to a large item in the expense of railway freighting. Proper facilities in this department, not only economize current expense, but greatly reduce the chances of error by the tallying checks. The station should be provided with suitable machinery, according to the value and extent of the traffic, so as to render available the economy of horse or mechanical power, the latter preferable where the business will warrant it. This is particularly important in a large grain trade, when the transit is between the railway and water craft; in which great improvement has been made within a few years past, and the time and expense of transit has been greatly reduced. I have known a station that, on a charge of one cent per bushel on grain, paid all the expenses, from the time the car was brought to the station until the grain was on shipboard, and, in addition to these expenses, paid a rent of 20 per cent. on the building and machinery; the work done, included the unloading the car, elevating the grain to a shipping loft, and weighing and conveying it to the hold of the vessel; the latter being done at the rate of 3,000 bushels per hour for a single hold to receive, and double if the vessel had two receiving holds to operate at the same time.

The floor of the building, or whatever constitutes the channel of transit from or to the railway, should be under cover, and lie between the vessel and the car, as this is most convenient for the transitory storage often required. For miscellaneous goods, one story high is sufficient, and to prevent exposure from fire, it is best to have no loft in such a building, which would prove of small account, as storage is not wanted for such length of time as would warrant the expense of hoisting: but a portion of the loft may in some cases be advantageously occupied as an office for current freight business. For goods not liable to theft, open sheds are economical and convenient, and to considerable extent may be used for housing freight.

In connection with the freight buildings, there should be spare tracks, on which trains of cars may be conveniently made up, so as to require as little movement as practicable, and prepared to start out on their appointed time; also, for incoming trains that may not be able at once to take their proper position for unloading; and for empty cars to stand, while not wanted, or waiting for small repairs. To provide the spare track, so that all these objects may be accomplished with the least labor and hazard of confusion, will be found, in most situations, to require much study. The entire freight arrangements at important, and especially at terminal stations, should, if possible, be kept wholly disconnected from the passenger station after branching from the main track.

At *way stations*, the extent of accommodations will depend on the amount of business, and somewhat on the competition of other railways; the latter may require more careful economy, and better inducements to shippers than would otherwise be necessary, and should be carefully regarded in the plans for station arrangements. They should be more simple than the terminal stations, as the business will be less; but so far as they have business, they should be arranged with similar views of economy and convenience. Where a large grain business is done, the loft is used for grain, and the main floor for miscellaneous goods. In many cases the grain is elevated to the loft by a roadway, on which the team that brings the grain to the railway travels up an easy grade; in others, an inclined-plane railway extends to the loft, and when the grain is unloaded from the wagon into a light car standing at the foot of the plane, the car is drawn up by a rope over a pulley—the rope being hitched to the wagon, and as the team moves off with the wagon, it draws up the grain car. In either case, the railway company (except for the fixtures) is at no expense for the elevating power; and either method is usually satisfactory to shippers. The latest and, as I consider, the most approved roadway, is carried up into the end of the building, and the team passes through and down by a bridge from the other end of the building. The load is weighed before it goes up, and the grain emptied directly into the garners, requiring no machinery to transfer it from the wagon—the whole is very economical. I have only seen this plan of roadway on the Chicago and Rock Island Railway. The elevated roadway is best for a large business, and the inclined railway for a small business. From the loft, the grain is conveniently loaded into the cars by spouts. The elevating is sometimes done by machinery, with horse or steam power; and in some instances this is required by local circumstances; but it is very expensive, and one of the other modes, above described, should be adopted if practicable.

Passenger Station.—The indispensable thing here is, to provide sufficient tracks to accommodate all the incoming and outgoing trains, so that they need not interfere with each other; and also such as may be required for standing coaches. The terminal stations should be covered, where there is much traffic, so far as to afford shelter for passengers and baggage, while making up and discharging trains. Waiting rooms should be provided for passengers, who will be more or less detained for the trains, and for conveyances to take them from the discharging train. A convenient ticket-office should be placed in this building; and rooms provided for baggage, especially such as may not be at once called for. The covered part, where the trains are made up and discharged, should have a floor over the whole area. Formerly it was the general practice to have a raised platform alongside of the out-going and in-coming tracks, elevated from one to three feet, to facilitate the passengers in getting in and out of the coaches; but this has generally given place, in modern stations, to the plan of making the whole floor on a level with the track, and which is no doubt more convenient and safe for general use; it is more convenient for receiving and discharging passengers, and if more than one track is required for out-going and in-coming trains, it is perfectly convenient to use either, as no track-pits are in the way to embarrass passengers, or expose them to hazard. A further advantage in a level floor arises from the convenience afforded of moving baggage trucks in any direction.

An office will be required at one of the terminal stations, where the general business of the company should be transacted. This office should be convenient to the station, as here all reports, way-bills, check-rolls, and accounts are sent for arrangement and disposal. If the situation be favorable, it may be combined with some other station building; but in regard to safety from fire, it is usually better to have a separate building, with ample fire-proof safes, as a large quantity of documents will be collected here in a few years, the loss of which would be very great to the railway company, and seriously embarrass the management.

The terminal stations will require some provision for engine and coach sheds; the latter may, in some cases, be provided for sufficiently in the passenger coach shed. Shop accommodations in connection with the engine-shed, should be provided to some extent for small repairs. In most cases, it will be best to have the main shop for repairs of engines, coaches, and cars, at one of the terminal stations. In the latter case, the main shop, with engine and other sheds, should be placed outside, and as near to the station as the ground will permit. If the situation of ground permit, it should have its side-tracks immediately back of the switches, or points that separate the passenger from the freight trains as they go into the station. That portion of the shop designed for repairs of engines should be in convenient position in reference to the engine shed, so as to require the least practicable movement of engines between the shop and shed. The coach and car-shop will require a paint-shop connected, for repainting and varnishing, and liberal spare track for standing-room. It will be economy to so arrange the shops, that the same stationary power will drive the machinery of both. The blacksmith's shop should be convenient to do the work of both the other shops.

An important feature in shops of this kind is, some arrangement by which an engine, coach or car may be placed in and removed from the shop without disturbing the repairs of others; this is often done in a very bungling way. The transfer-table, if well made, so as to work easy, is, I think, the best method; when properly constructed, it works easy, and meets the wants of the case in a most direct and convenient manner.

The magnitude of the various station structures must depend on the traffic to be provided for. This is likely to be larger at some future day than for the first few years of the operations of the railway, and it will not, in most cases, be economy to make the outlay in the commencement sufficient for the probable future wants of the railway; and with this view, the plans, so far as may be practicable, should be made with reference to enlargement, as further accommodation may be demanded by increasing traffic; and in most respects this may easily be done, if seasonably looked after.

It often happens that railways are put in operation before any considerable progress is made in station buildings; but if few or no buildings are prepared at the opening, the engineer should have his plans matured, so that the work, as it may progress, shall be in conformity; that the whole, when completed, may have the order and symmetry that will secure the greatest ultimate economy in conducting the traffic of the station. I am well aware that the engineer of a railway is not always responsible for ill advised station arrangements, especially when not completed until after the railway is put in operation; for then a superintendent of the operating department, if not of the railway in full, is put in charge, and in some cases, he may have little education to fit him for studying the requirements of a railway station; but usually enough of self-reliance to regard his capacity for such matters as much beyond that of the engineer; and if he has, as is most likely, the influence of the board of managers, he will often disregard the engineer, and under plea of some present necessity, proceed with some part of the works with little regard to interference with any well devised arrangement that the ultimate wants of the station may demand. If the engineer is incompetent, a competent one should be obtained, as his education is best designed to secure the capacity wanted in such matters; nor is it the proper province of a superintendent to control him. A further difficulty is sometimes experienced by agreements between the managers and citizens of towns in regard to station arrangements, that prove very embarrassing to the future business of the railway. In one of these ways, or in the failure of the engineer to prepare suitable plans, may be traced the awkward and inconvenient station arrangements that are not unfrequently to be seen on railways.

CHAPTER XIII.

LOCOMOTIVE ENGINES.

The locomotive engines are a part of the rolling stock, and constitute a portion of the construction of a railway, and should be carefully considered by the engineer.

After an extensive use for over a quarter of a century, the locomotive engine is a machine that does not fail to inspire the feeling of wonder and admiration in those who witness its movements, and especially those who are familiar with its mechanism, and can appreciate its duties and its power. That such a machine may travel in day-light—in the darkness of midnight—in sunshine and in storm—through cultivated fields and dense forests—through hill and over dale—at the rate of fifty miles per hour—carrying hundreds of people, as it thunders its way onward to the distant station—and in all its marvellous speed and power, holds fast to the narrow track, is truly a wonder; and though a thousand times witnessed, I never fail to admire and enjoy, as it passes with its long train, this splendid specimen of art.

Locomotives should be adapted to the work they are required to perform. As has been noticed, they are now made much heavier than when the system was first brought into use. It should be kept in mind that heavy engines require more expense for repairs of track, and for their own repairs, than lighter ones. A judicious arrangement in the weight of engines will have great influence on the economy of working the railway, and it should not be forgotten, that the object of the proprietors is, remuneration for the outlay they are required to make.

In the service of all railways there is a considerable portion that may be done most economically by light engines; and on many railways a large proportion. The speed required will have much influence on the weight of engines. No intelligent railway manager will run at a higher speed for express trains, than 25 miles per hour, including stops, unless compelled by some competitor for the passenger traffic. Attempts have been made to show the relative cost of high and low speed; but none, so far as I know, on any well-founded data. It is not easy to get at a basis that would be sufficient to warrant any computation on which a rule or formula to determine this question might be made. Fast trains and slow trains—heavy engines and light, are used indiscriminately on the same track, and it is impossible to decide how much of the wear of track is due to each. The wear and consequent repair of engines may be determined from the careful accounts of work done and repairs made on each class, and the relative cost and work may be ascertained with satisfactory approximation. Beyond this, so far as I know, this question of comparison in all its bearings can only be estimated from the statistics of mileage, and long and careful observation of the influence of weight and speed on the track, the rolling machinery, and the hazard of accidents that necessarily follow. In regard to the latter, their number must be increased by high speed; also their intensity as to damage to persons and property. As before observed, we have no *exact* data to determine the relative cost of different rates of speed: some have estimated it to be as the squares of the velocity; others a lower rate of difference; my own opinion is, that on the whole the former is not far wrong. With two engines of suitable weight, one to haul a train at the rate of twenty miles per hour, the other to take the same train thirty miles per hour, the cost per mile for the train having a speed of twenty miles per hour may be approximately estimated at one-half the cost of a train having a speed of thirty miles per hour. This of course embraces not only the engines, but the cars, coaches and track, all of which suffer increased wear and damage from increased speed.

At the celebrated trial of engines, on the opening of the Liverpool and Manchester Railway, the successful competitor, the Rocket, was mounted on four wheels; one pair driving wheels and one pair supporting wheels. This plan of four-wheeled engines, with one pair of drivers, or with both pairs connected as drivers, continued in general use for several years in England: it was used on the railways of this country, mostly in the eastern States, until after the opening of the western railway of Massachusetts; and down to 1845, was more or less used in that section. In 1831, the Mohawk and Hudson Railway Company imported an engine from England, made under the direction of George Stephenson, the distinguished engineer in this department of the profession. This engine was on four wheels, all drivers, and weighed about *seven tons.* The wheels were four feet diameter, and the axles four and a half feet apart, or from centre to centre. The performance of the engine was at that time satisfactory as to power. The frame was twelve feet long, and the axles, being four and a half feet apart, it projected beyond the bearing on the axles near four feet each way. It was readily observed that a vertical inequality in the surface of the rails, caused a vertical motion at the ends of the frame of about double this inequality, producing an unsteady and shaking motion to the frame of the engine, very unfavorable to the machinery and the engine-men. It was further evident that this leverage action of the frame was unfavorable to the track. The first thought for a remedy for this difficulty was to spread the axles further apart; but to do this to such an extent as would materially remedy the evil, was at that time considered inadmissible, on account of the increased labor and danger it would cause in passing curves in the line of railway. The track of the Mohawk and Hudson company was a well constructed one of the kind, very direct, and in good order, probably as smooth at the time referred to as any railway; it was a flat bar or plate rail, laid on southern pine rail timbers well secured; still, the action before mentioned was very unsatisfactory.

My observations on the action of the Mohawk and Hudson engine, led me to inquire into some means of providing a remedy. There had been six-wheeled engines put in operation, but they were on a single frame, the third pair of wheels merely added for support, and all worked in the single rigid frame; at that time no six-wheeled engine, or of more than six wheels, had been successfully run at high speed. It appeared important to provide guiding wheels that should be geared favorably to follow the track, and support one end of the engine frame, so that the engine and all its working parts would be supported by the same rigid frame, as on the four-wheeled plan. While engaged with these considerations, the attempt was made by a fellow-engineer, to mount an engine on eight wheels, geared as two wagons, so coupled that each would be free to conform to the curves of the rail, and the machinery to conform to the changing parallels of the two wagons by movable joints. Here was the idea of working two wagons, so coupled as to constitute an eight-wheeled vehicle, but without a common rigid frame, and consequently the machinery, resting on two separate frames, depended on their movable joints to adapt them to the changes of parallelism constantly taking place on the rail. A similar effort had previously been made, to adapt two wagons as support for an eight-wheeled engine; but the plan had not succeeded in a manner to be practically useful. It did not appear to me that any plan would succeed, that did not provide a rigid frame for the engine machinery. The difficulty appeared to be in obtaining a connection between two frames, that should work free and be secure on the rail under high speed. There was no doubt it would work well at low speed; but nothing of the kind had been adapted to, or previously attempted for high speed, then much demanded for railway travel. Two four-wheeled cars, or wagons, had been coupled together, so as to form one eight-wheeled car, for transporting long timber and heavy stone, but it gave no confidence for high speed.

After devoting a good deal of labor to this subject, I prepared a plan, which it is hardly necessary to describe at this time, as it is in general use in the locomotives of this country. This plan, in its general features, had a guiding truck, or a four-wheeled car, arranged as best adapted for following curves on the rail, and keeping on the track, and, at the same time, supporting steadily the forward end of the engine frame. The plan of the engine was prepared in the fall of 1831, and sent to the West-Point Foundry Association, who built the engine, and it was placed on the track of the Mohawk and Hudson Railway in the summer of 1832. The working of this engine (named the Brother

Jonathan) satisfied me that the truck principle would be successful, though the engine was not so in other respects, the attempt having been made to adapt the boiler to the use of anthracite coal, and this required to be changed, which was done the following winter. I then prepared a new plan for an engine for the Saratoga and Schenectady Railway, following substantially the same plan, except as to the boiler, and sent it to George Stephenson, Esq., of Liverpool, who constructed the engine, and it was placed on the Saratoga and Shenectady Railway early in the following summer (1833).

As this truck system has come into general use in this country (it was never patented), it may be interesting to refer to its early history, as put on record at the time, and I therefore think I may be excused in making the following quotation from the "American Railroad Journal" (1833), vol. ii. p. 468.

LOCOMOTIVE STEAM ENGINE. BY JOHN B. JERVIS.

To the Editor of the American Railroad Journal.

Dear Sir: The locomotive steam engine for the Saratoga and Schenectady Railroad, of which I promised to give you some account, was put on the road the 2d inst., and has been in regular operation since, making usually two trips per day each way, equal 84 miles, and carrying daily over the road about 300 passengers.

The engine was made by George Stephenson & Co., at Newcastle, England. The boiler has tub lar flues, on the same plan as all of recent construction at that establishment. The leading object I had in view, in the general arrangement of the plan of this engine, did not contemplate any improvement in power, over those heretofore constructed by G. Stephenson & Co.; but to make an engine that would be better adapted to railroads of less strength than were then in use in England; that would travel with more ease to itself, and to the rail on curved roads—and would be less affected by inequalities in the rail, than is attained by the arrangement by the most approved engines.

You are aware of the fact, that the Saratoga and Schenectady rail is constructed with timber, capped with an iron plate. This kind of road cannot be expected to bear as heavy weight on the wheels of its carriages as those that have an entire iron rail; and, in order to obtain that degree of power which is desirable for an engine intended for high speed, it becomes an object to put the weight on six wheels, instead of four. Engines mounted on six wheels were constructed several years ago in England. The object was to distribute the weight on more points, to make them easier for the road than the four-wheeled engines; for even with the iron rail, the heavier carriage is injurious to the road. There was difficulty, however, in the practical operation of the plan adopted. The load was found to bear at times very unequally on different wheels, owing to the inequalities in the road; and having all their wheels under one frame, they did not work as well on curved roads as the four-wheeled engines, which could be geared much shorter. In consequence mainly of these difficulties, the six-wheeled engines were abandoned, and I believe no attempt has since been made in England to use more than four wheels.*

In the Saratoga engine, I have adopted two distinct frames. One

* Since the above was written, six-wheeled engines have been mostly adopted in England and other parts of Europe, and will be noticed hereafter.

frame embraces four wheels in the same manner as a common wagon: these wheels are small (32 inches) in diameter and of uniform size; one end of the second frame is mounted on the third pair of wheels, which are the working wheels, and the other end is rested on friction rollers, in the centre of the first frame, to which it is secured by a strong centre pin. The small wheels, with their frame, work on the road the same as an independent wagon; and being geared short, they go round a curve with as much ease as a common wagon, and being leaders, they bring round the working wheels, and the large frame on which the whole machinery of the engine rests, with as much ease as practicable. By this method it will be seen the engine may pass a curve with the same ease as a common railroad carriage, having the same weight on the wheels. The machinery of the engine is not affected by the curve motion of the carriage. In order to give the four-wheeled engine carriage as much facility as practicable in turning curves, the wheels have generally been placed near together, bringing the bearing points of the frame so near the centre, in a longitudinal direction, as to cause the inequalities of the rail to produce increased motion at the ends of the frame, and consequently to the engine and boiler which is connected with it. This, in the English engine belonging to the Mohawk and Hudson company, was such as to make the motion very unfavorable to the engine, and severe on the road. By allowing the bearing points to be near the ends of the large frame, and resting one of these points on the centre of the small frame, as is done in the Saratoga engine, the difficulty is almost entirely remedied.

The engine was set up at the shop of the Mohawk and Hudson Railroad Company, under the direction of Mr. Asa Whitney, the present superintendent of that road, and who has from the commencement, had charge of the machine-shop connected with it.

Thus far, the engine appears to do all that was anticipated from

No test has yet been made of its power; but, from the rapidity with which it generates steam, there appears no doubt of its performing all that it was calculated to do. It passes a curve without any more appearance of labor than a well-geared common carriage. The principle of its arrangement does not admit of more strain coming on any one wheel than is assigned for its regular labor. The motion of the engine is highly satisfactory; it moves with almost as smooth and steady motion as a stationary engine; it travels over the road in an elegant and graceful style.

I made a plan for a six-wheeled engine for the Mohawk and Hudson road, which was completed and put in operation before I made the plan for the Saratoga engine. This engine proved satisfactory so far as regarded the principle of a six-wheeled carriage, and was an important pioneer for the second plan. The superior ease with which this engine moved, both for its own machinery and the road, led to the determination to alter the English engine on the Mohawk and Hudson road, so that it could be placed on a six-wheeled carriage. As the engine was particularly arranged for four wheels, this could not conveniently be done in any other way than by communicating the power through the intervention of a bell crank,* which was very successfully done by Mr. Whitney. This engine is now working on six wheels, and the ease and smoothness of her motion, over that she had when on four wheels, is very striking.

The arrangement on six wheels does not admit of the wheels under the main frame being connected with those under the small frame; consequently, we can only obtain the adhesion of one pair of wheels. This, however, is hardly of any importance where high speed is wanted.

Should further experience confirm what the operations thus far

* An improved method has since been adopted, and will be discussed hereafter.

appear to warrant, the plan of the Saratoga engine may be viewed as a valuable improvement. She has used for fuel a coke of inferior quality, made in New York, with which she has worked very well.

Yours, etc.,

JOHN B. JERVIS.

ALBANY, 18th *July*, 1833.

From what has been stated in the preceding pages, it will be seen that the truck arrangement was adapted to an engine with only one pair of driving wheels. At that time, passenger engines were mostly made with one pair of drivers, even for four-wheeled engines, both in this country and in Europe; and the practice has more or less prevailed, to within ten years of the present time, and is still, I think, the prevalent plan of English passenger engines. There can be no doubt that a single pair of drivers work more kindly, easier for the rail and the engine, and will do more work for its power, than any others. To give a better support, and to relieve the weight on the driving wheels, a pair of trailing wheels, usually termed relief wheels, were placed behind the drivers, receiving a portion of the weight of the fire-box. As engines were enlarged, this method of support was the first measure adopted in England to obtain for modern engines the benefit of six wheels. But as the want of increased power grew with the development of railway traffic, more drivers were wanted to give the necessary adhesion to the rail. This was more especially required for freight engines, and to a greater or less extent, for passenger engines. It did not require much time to adapt the arrangement to two pair of drivers, and to some extent three pair have been adopted, and used with favor by some railway managers. Still the truck is generally, and almost invariably used as the leading wheels, supporting the forward end of the engine frame. Comparatively a very small number of freight engines are made with three pair of drivers, without a leading truck, and those that have been tried, as far as I know, are but little esteemed; they are very severe on the track, and subject to heavy repairs.

Some of the truck engines were sent to England; but were not received with favor. When in that country, in 1850, I endeavored to ascertain the objection to them; and the only one I was able to learn, was an impression that they would not keep the track well under high speed. If this had been a good objection there, it certainly would have been here; but no such difficulty has been experienced with us. We have tried the principle thoroughly, and in fact carried it to extremes, in the close gearing of the truck, barely giving clearance to the wheels in the line of track; which is the worst position that can be given to the wheels, for safety in keeping the track. A truck, or what is the same thing, a railway wagon on four wheels, should not have its two pair of wheels either too close together, or too far apart; in either case it is more liable to leave the track; it does not work easy when too long, and is unsteady when too short. It is better for the track, and also for steadiness of motion, to have the truck wheels as far apart as is consistent with easy movement on the track; and I have never seen a good reason for the fashion that prevailed very much a few years since, of very close gearing of the truck. Four and a-half to five feet between centres of axles, may be regarded as securing a good result. But notwithstanding errors in close gearing, nothing has occurred in this country to impair confidence in the safety of the truck engine for keeping the track.

In regard to locomotive engines in the railway practice of Europe, the English plan, and to a great extent the English manufacture, has been adopted. They found the four-wheeled engines insufficient for the enlarging demands of traffic; and soon adopted, as before noted, the third pair, called trailing wheels; the leading and the trailing wheels were small, and the drivers (one pair) were placed between them. As before observed, the engines, in the early history of railways (reference being had to the time subsequent to the trial on the Liverpool and Manchester Railway), had generally but one pair of drivers; and this was continued as the general practice for passenger engines in England down to 1850, at which time very few had two pair of drivers; and, as before stated, this is believed to be the prevalent practice at this time. The freight traffic required more adhesion, and for this service, the trailing wheels gave place to a pair of drivers. In all these arrangements on English engines, the six-wheeled engines had all their wheels attached to one frame, and worked on the track in the same manner as the four-wheeled, except that the frame was longer, and the bearing of the front and rear wheels nearer the ends of the frame. It will be perceived that no ease in the working on the track was gained by this, over the four-wheeled plan; it is, in fact, so far as the movement of the carriage part of the engine is affected, the same as to extend the four wheels to the same distance apart as the leading and trailing wheels are placed, and the third pair, whether as trailing or driving, are merely so much additional support to the engine. It is, therefore, apparent that the English six wheeled engine has all the objections of a long-geared railway wagon in passing curves. To remedy the oscillating motion, both lateral and vertical, the front and rear wheels have been spread from four and a half feet, to ten, twelve, and even fourteen feet apart. On a railway nearly straight, there would be no objection to this; but railways are not all straight, either in Europe or America, and all must have turn-outs from one track to another, that necessarily require pretty sharp, if not very sharp curves. In England they have been more liberal in expenditure to secure a good direction of line, than we have in this country; but the more favorable formation of surface with us, has allowed us to make lines that, on the whole, will compare favorably with theirs. The superior ease in the movement in the truck engine on curved rails, over that of the long-geared, rigid frame of the English engine, is so manifest to an intelligent observer, that it seems strange that European engineers should not appreciate it, and that they should continue to discard it after the experience of this country has so long given unequivocal preference to its use, on three times the length of railway they have in England. I saw a few truck engines in Belgium, in 1850, and at this time, I believe, a large proportion of those used in Russia are of this kind. That the English should at first have entertained the apprehension that the truck would be liable to leave the track, did not surprise me, for this was the most serious apprehension in my own mind when engaged on the plan, at a time when no such machine, or any double car or wagon, had been successfully used for any purpose but such as admitted slow speed—as in transporting long timber or other heavy material for short distances. But more than twenty years of experience, and most of the time on thousands of miles of railway in this country, would appear to afford experience enough to satisfy them of the groundless nature of their fears. Perhaps they do not respect our practice as much we do theirs, and hence our advantage.

In all that relates to thorough building, and in avoiding all tinsel in their engines, the English are our superiors. It would be much better for us, if we substituted wrought iron for much of our cast iron material, and left off the brass ornaments, which are not in sound taste on a machine that is made to perform work, and not for show. An English engine is black, with very little bright work, except where it is necessary to reduce friction in the moving parts.

I have been somewhat particular in my remarks on the truck engine. It is proper that I should state frankly that I have two objects for this: first, the truck has proved to be a highly important improvement in the locomotive engine—is of American origin, and the credit is due to American engineering. In this view it is interesting to all who have a taste for inquiring into the origin of valuable improvements, and who desire a reliable history of them; and secondly, it is due to myself that the public should understand that I was the inventor, and had the sole responsibility of introducing this improvement in railway machinery. As I took no patent for the invention, that kind of publicity has not been enjoyed; and very few persons who now use the engine, have the least idea to whom the railway public is indebted for its introduction. I do not complain; as an American engineer, it was a satisfactory compensation to be able to present to the railway interest a valuable improvement, and to see it rapidly dissipate in this country the preference that, in many quarters, had been tenaciously given to the English plan.

It will be seen from the preceding remarks that the truck engine is materially different from the English, and in our country it so completely takes the precedence on nearly thirty thousand miles of railway, as to lead to the belief that it will eventually do the same in Europe.

It has been stated that a single pair of driving wheels work more favorably to the ease of the engine than two pair; and so far as the adhesion can be obtained without too great weight on one pair, it is the best engine, especially for high speed. The coupling of two pair of drivers increases the number of parts, and the slightest difference in the diameters of the two pair coupled together, causes severe strain on the working parts of the engine and increases the wear on the rail. The English continue this plan up to a weight of five and six tons on a wheel, which not only requires a strong rail to support it, but is attended with severe wear both on rail and engine. My opinion is, that one pair of drivers with three-fourths the weight of two, will do the work of two pair for high speed, with more ease to the engine and less wear on the track. In this country, at the present day, an engine with one pair of drivers is rarely seen. For passenger service, if the work required can be done with an engine having about six tons on one pair, or three tons on a wheel, it will be the best economy to adopt one pair; and on railways having a large passenger traffic, requiring high speed, I should prefer four tons on a wheel with one pair, to three tons on a wheel with two pair.

Passenger engines designed for high speed, where t may be important to secure punctuality in the time, in their connections with other railways forming portions of a general line, must be provided with power to meet difficulties beyond what would be otherwise necessary; such as strong head winds, an unfavorable state of the rail, and to make up time that may be lost by incidental delays. The importance of punctuality in these connections depends mainly on the position of a rival line, and, if often missed, the reputation of the line will suffer, and the importance in this respect will depend on the value of the traffic in competition. Rivalry is a feature in railway affairs that tends greatly to increase expenses, and no doubt it often happens in these struggles that the traffic secured costs more than it is worth. The local traffic of a railway may be done with more definite calculations for economy; there is no necessity of running power in excess, and often to great waste, as is frequently the case in rivalry with other lines; and hence, so far as regards the local traffic, or any other that is not affected by competition, the power of the engine may be economically arranged to provide for it.

Freight engines should be designed to travel slowly, and to these the objection to connected drivers is not so important, and the necessity of greater adhesion will generally determine in favor of that plan. The weight and power of freight engines will depend on the extent of the traffic. On a double track, they may be arranged so as to make the most economical transportation; for in this case the number of trains may be increased to meet almost any amount of traffic; but on a single track railway there may be a difficulty in sending as many trains as the freight may require, and larger trains than otherwise desirable must be made up, and of course an engine of larger power, or duplicate engine, must be employed to haul it. The latter plan is often adopted, either by dividing the train into sections, or by coupling two engines together; each of these methods has its advocates, while neither is much liked, the preference being generally given to one engine of sufficient power to do the work alone, if that be practicable. The latter mode is certainly the most simple, while its economy in many cases is more than doubtful. An engine best adapted for a freight train, with a judicious time table, will, in nine cases out of ten, haul the train, and the inconvenience of occasionally putting the train in two or more sections, to meet an extra run of freight, is a trifle compared to the injury of an engine that must always take the train, whether large or small. So long, however, as single track railways prevail, there will be an excuse for a class of larger freight engines than is profitable for a railway. The question for the engineer is, to consider well the character of the railway and the traffic it is to carry, and design the engines of all classes on the plan that will conduct the anticipated service with the greatest net profit. In this the engineer will bear in mind the wear and repairs of track, as well as that of the engines and will find it a field requiring his best judgment and experience.

In this country wood has been the usual fuel for locomotive engines. Coke has mostly been used in England,, and is a much more convenient fuel than wood; it does not require, as the wood-burning engines do, a screen or spark arrester, to prevent the escape of small burning coals or cinders, carried through the smoke-pipe from wood. This spark arrester diminishes the power of the engine, and is a heavy appendage to the machine. Coke is a very convenient fuel; it is of less weight and much less bulk than wood, per mile run. They have been compelled to use coke in England, not being allowed to use coal, which is a much cheaper fuel, on account of the smoke, which seems to be a very offensive thing to an Englishman: but recently, it is said, a mode of consuming the smoke, by introducing a volume of atmospheric air, so as to unite with the gases and smoke as they rise from the surface of the burning coal, causes so complete a combustion, that no smoke of consequence escapes from the smoke-pipe. If this shall be found by experience to succeed, it will greatly improve the economy of fuel in English engines, and in all others where bituminous coal of good quality can be cheaply had. Coal has been used in this country for several years in freight engines, and to some extent, recently, in passenger engines; the question must be one of increasing importance, as wood becomes scarce and dear, which in some parts is already the case. Coal, however, in most parts of this country, is dearer than in England. Even the anthracite is more expensive, and good bituminous coal, suitable for locomotive engines, can only be had east of the Alleghany range by the cost of long transportation. Still, there are many railways on the Atlantic district, that will find bituminous coal to be cheaper than wood. There have been several attempts to use anthracite coal in locomotive engines, and I believe it to have succeeded very well in some cases, for low speed, as required for freight trains.

In the western States there is a large supply of bituminous coal, some of which is of very good quality, but a large portion is inferior and will only be used from necessity; and this necessity will, no doubt, bring much of it into use in the prairie districts, where wood is scarce; and the engines must be adapted to use the fuel that can be obtained with the greatest economy. There are many railways in this country situated so that they can purchase good wood, delivered at their stations for $1 50 to $2 50 per cord; in such cases, coal will not supersede wood, except in those localities that furnish a good article at a low price. The engineer will see that the question of fuel to be used in engines, will depend on the price and relative quality at which the different kinds can be obtained, for the locality to be provided for.

CHAPTER XIV.

COACHES AND CARS.

THE trains used in England for railway vehicles are, namely, *wagons*, for such as are adapted to freight, and *coaches*, for those adapted to passengers. In this country the term *car* has generally been adopted for all railway vehicles, though to some extent the term coach has been used for passenger cars. The term car has the advantage of brevity, and of signifying a vehicle adapted to a railway, as contradistinguished from the vehicles of ordinary roads. If we adopt the term coach, for passengers, it must have the prefix of *railway*, to distinguish it from coaches used on common roads, so we do not abbreviate or cut off the prefix by this term; the prefix is necessary in either case, to hold the distinction between vehicles adapted to freight, and those for passengers on the railway. If the term coach be adopted, then to make the distinction complete, it must be railway coach; and in the other, or American term, passenger car; between these there is little to choose for brevity in writing, and one may be considered as easy of pronunciation as the other. In the ordinary intercourse of the operations, in using the term coach for passenger, the railway prefix would be omitted, whereas, in the term car, the prefix of passenger must be used; the advantage, therefore, of brevity and convenience in the parlance of the station, is in favor of coach. The term car, if applied exclusively to freight vehicles, has the advantage of sufficient distinction from other railway vehicles, and from those on common roads. If, therefore, the term coach is applied to vehicles for the conveyance of passengers, and the term car to those for freight, baggage, etc., we shall secure terms, expressive, and of the greatest brevity and convenience for the use of the operatives on the railway, and shall lose nothing in writing. The full terms will then be, railway coach, freight car, baggage car, platform car, cattle car, gravel car, and hand car. All, except the coach, are in general use in this country, and as before observed, to some extent, this is also used. These terms will be adopted in the following remarks on this branch of the subject, as best calculated to serve general convenience, and not inconsistent with usage.

Railway coaches and cars were formerly four-wheeled vehicles, and this was the general practice down to 1838. Shortly previous to this date, the plan of eight-wheeled coaches and cars began to be introduced in this country, and they are now in general use for both freight and passenger traffic. It will require no labor to show that the plan was simply a duplication of the truck of the locomotive engine, then, and for several years previous, in successful operation. The truck that would work successfully for the leaders of an engine, could not fail to work for a car; and if it worked for leaders, there could be no doubt it would work equally well for the hind, or trailing end of the car. In this way two trucks were adapted to take the place of two pair of wheels, supporting the upper frame at both ends, in the same way as the truck supported the forward end of the engine. Such a coach or car, it was easy to see, would run equally well in either direction. The frame of the coach or car, resting on the centre of the truck, would secure to its motion the same advantage of steadiness, that it gave the engine, and the same facilities for easy and safe working on curve lines of rail. From its action, as described on engines, it will be perceived, that the truck may be made so as to pass curves with the greatest ease and safety that is attainable. The ends of a four-wheeled car, extending about as much from the bearing point on the axle, as the axles were from each other, produced at the ends of the frame, double the amount of vertical and horizontal vibration that occurred at the wheels, or bearing point on the rail; whereas, in the truck arrangement, the motion of the main frame at the point where it rests on the centre of the truck (near the end of the main frame) is only half as much as that due to the motion of the truck wheels, thereby reducing the horizontal and vertical motions of the end of the frame to a trifle over one quarter of what they were in the four-wheeled arrangement.

The four-wheeled coach was made for the seating of eighteen to twenty-four passengers. The eight-wheeled coach was originally made for seating thirty-six to forty persons, and has been increased in size to provide for seventy-two; but the general rule at this time is to provide for sixty seats. Coaches have been made with six-wheeled trucks, and some of this kind are now in use, but they are not generally regarded with favor, and very few railway managers adopt them. The four-wheeled truck is more simple, has less friction, and meets every requirement that is desirable. The only thing gained by the six-wheeled truck, is more wheels, and the capacity for a larger coach, which I do not consider economical in the working of the railway. The four-wheeled truck is a very simple machine, and is adapted to produce the best motion, with the least friction, and admits the form that will traverse curve rails with the greatest ease and safety. It admits as large a coach as is wanted, with the least number and weight of wheels required to secure its peculiar advantagss; and may be made to reduce the pressure on the rail, by reducing the coach to the same capacity as the four-wheeled coach.

Large coaches are regarded by some persons as more safe for passengers than small ones. No doubt a strong coach is safer than a weak one; and a large coach is necessarily weaker than a small one, if its weight is in the simple proportion of the relative sizes. To make a large coach as safe as a small one, its weight and strength must be in geometrical instead of direct proportion. The danger of injury from accidents, arises mostly from the momentum of the train, and the question is, what size of coaches in the train will experience a sudden check to the momentum the most severely. If there was but one coach in the train, and that of the size of five ordinary coaches, the momentum would all be in that one coach, and the whole action would be instantaneous; but if the train was in five coaches, they would all be separated a few inches, and the first coach, at the instant of meeting an obstruction, would only act by its own momentum, and the second, as it came up, would first act on its buffer springs, which would modify its force, and between the instant of time and the action of the buffer springs, the momentum would be somewhat checked; and so to the last coach, the joints, the small separation, and the action of the buffer springs, would reduce the momentum of the train by dividing its power. It is, therefore, obvious that small coaches tend to check or to ameliorate the force of momentum, and thereby reduce the severity of its action and the danger to the train.

When the eight-wheeled coaches were first introduced, it was believed, that in cases of accident, they were not as liable to overset as the small four-wheeled coaches, which I think was true at that time; but as speed has increased, it is quite common for them to be upset, in accidents by which they are thrown from the track, and the roof and floor badly broken up. If the coach, large or small, when thrown from the track, is strong enough to resist breaking up, there is not usually much personal injury to passengers. There is, doubtless, a medium for the size of coach that is most safe, and I think a forty-seated coach is as good, and quite as likely to be safe as any larger size.

In England, and generally in Europe, the managers adhere to the single frame in coaches and cars, not having adopted the double frame, or truck plan. In their railway coaches they have generally spread their wheels further apart, so as to reduce the motion at the ends. It will appear obvious, that the close gearing of the wheels in coaches, produced an action at the ends of the frame, similar to that which has been described as unfavorable in locomotive engines. The coach wheels were originally about four and a half feet apart from centres, in a longitudinal direction. This distance was gradually extended until it reached ten feet, in four-wheeled coaches; and finally, the third pair of wheels was introduced, and the extreme wheels spread to twelve and fourteen feet. This plan materially reduced the motion at the ends of the coach; but it is subject to the objection of increased friction on the rail, and cannot be as safe in its adherence to the track, on the curved parts of the rail. For reasons before stated in regard to engines, the end motion of a single frame must be greater than when the main frame rests on the centre of a truck frame; and if the English coach had no auxiliary means to steady its motion, it would be a much worse riding coach than the American; but the English have adopted a very good plan to steady the coach, by means of a right and left coupling-screw, which draws the coaches to a close bearing on their excellent buffer springs, which greatly controls the vibratory motion caused by the inequalities of the rail. The passenger on a fast train very soon discovers whether or not the buffers are properly screwed up; in fact, the motion would be extremely unpleasant if the coach was left to act without any means of this sort to control the vibratory motion. With all the aid of this device, this motion, at points where the rail is not in good line, or when entering on a curve, is very sharp and unpleasant, even on their best lines of railway. If the same excellence was given to the springs of the American railway coach (both bearing and buffer) that is given to the English, the former would ride much more pleasantly than the latter. The decided superiority of the English springs go far to make up for the advantages the American coach derives from its truck arrangement.

The English railway coach with six wheels has four apartments for passengers, with six seats each for first-class coaches, and eight seats for second-class coaches, or respectively 24 and 32 passengers for each coach. The body of the first-class coach is about 22 feet long, and of the second class coach about 20 feet long; or the latter is about half the length of the American sixty-seated coach. It does not appear from the experience of English and American railway travelling that any increased safety is secured to passengers by large coaches.

On the continent of Europe, American railway coaches on the truck principle have been introduced to a limited extent: I believe they are mostly used on the Russian railways, and very few in other parts of Europe. In this country they have superseded (except for the horse railways of cities) all others, and undoubtedly possess advantages superior to any other now in use. But in their detail there is room for improvement. The truck arrangement undoubtedly adds something to the weight of the coach, as there must be the truck frames, which are an addition to the main frame; but this disadvantage is more than compensated by its superior ease of motion and traction, and its ease to the rail and machinery; and considering the facility of increasing the number of wheels, it is quite practicable to make a coach for a given number of passengers on the truck plan, with less weight on each wheel than on the old plan; and as railway science improves, and the influence of the weight on the wheel comes to be duly considered, the truck arrangement will be found to possess great advantages in the economy of railway management.

As before mentioned, the English railway coach spring is excellent; the American spring is far inferior to it. The latter is too short to afford the soft motion that is desirable both for the comfort of the passengers and the durability of the coach and the track. This has arisen in part from the erroneous short gearing of the truck, not allowing room for the proper length of spring. The short gearing was an early error, growing out of the supposed necessity of running most favorably on curved rails, which is gradually giving way to a better rule. With a truck geared with its wheels five and a half feet from centres, the spring may be much improved, and the truck made safe and steady.

We have gone largely into india rubber as a material for railway coach-springs; a material better calculated for a bouncing motion than for the legitimate wants of a coach-spring, and as the practice is now getting out of favor, it is hoped it may be soon laid aside, and steel, the best article now known, be wholly substituted. There are several points in which a soft, easy spring is important in coaches run at high speed: First, it is more easy and agreeable to the passenger; secondly, it reduces the wear and tear of the coach; and, thirdly, it saves all its cost in the wear and tear of track. The weight of the railway coach has gone up from

14,000 pounds to 21,000 pounds, and this great weight at high speed renders it highly important to secure the best springs that can reasonably be made.

As before stated, the American railway coach is usually designed for 60 passengers—the English taking their second-class six-wheeled coach for 32 passengers. The former carry seven and a half passengers on a wheel, and the latter five and a half passengers on a wheel. Comparing the American with the first-class English coach, the former carry seven and a half and the latter four passengers on a wheel. The American has cast iron wheels and the English wrought iron wheels. It has been observed, that the American coach, owing to its truck-frame, has greater elements of weight than the English coach; but the latter is notwithstanding about the same weight per seat as the former; and being shorter, is much stronger and better able to resist crushing in a collision. While the English coach is about the same weight per passenger as the American, the wheel of the first-class English coach carries *four* passengers, and the wheel of the American coach carries *seven and a half* passengers, or nearly double the number on a cast iron wheel than the English coach carries on a wrought iron wheel of about equal weight. Cast iron is not used in English coaches in any part requiring strength; even the jaws on the frame for steadying the bearing springs are of wrought iron, instead of cast iron, as in ours. Wrought iron, if it fail, is not like cast iron, liable to sudden rupture, but usually gives warning sufficient to guard against actual failure, and is therefore a much safer material. As evidence of the anxiety in this respect of the managers of American railways, it may often be noticed at stopping-places that men pass round the train with hammers, sounding the wheels in order to discover if they have received on the journey any checks not discoverable by the eye; a practice that cannot be neglected with prudence. Now, a material that requires so much watching should not be used for high speed if a better or safer can be had at reasonable cost—if indeed at any cost. But it is doubtful if cast iron wheels are any cheaper, except in first cost; and this extra first cost, it is believed, is compensated by superior durability; and if the damage by accidents caused by the breaking of cast iron wheels be taken into account, there can be no doubt the wrought iron is the cheapest wheel in the end, and far more satisfactory as to safety.

The American railway coach has no cross partitions, and generally no doors in the sides; it has a passage longitudinally through the centre through which the passengers pass to the seats on either side, and thus requires the body of the coach to be of an extra width equal to the width of the passage, which is an element of weakness, especially as it can have no cross support between the ends, except what it has in the roof and floor. There being no doors in the sides, adds to the strength in this part, and goes to compensate for the want of cross partitions, and if the coach was not unreasonably long and wide, would balance the want of cross support. The English coach is divided by cross partitions, inclosing two rows of cross seats in each division or apartment, which gives good facility for combining the floor and roof, but has the weakening effect of side doors, and if the body was very long, like the American coach, this would be a serious objection to its strength. But as it is only half as long as the American, and the body supported at three points instead of two, it must on the whole be regarded for equal material a stronger coach, and less liable to break up by collision. It cannot, however, be recommended to put partitions and their necessary appendages, side doors, in the long American coach; the strength of this coach is very dependent on its sides, and if the long beams were not supported by the deep side plank under the windows, they would require to be very heavy and well trussed with iron. It has been noticed that the English coach is about the same weight per passenger, while its plan has less elements of weight than the American; and that the former carries but little more than half the weight of passenger per wheel that is carried by the latter; that the former use wrought iron as a material where the latter use cast iron. These circumstances, with the general superiority of their track and police, sufficiently account for the greater safety of English railway travelling.

The plan of the American railway coach is adapted to at least an equal degree of safety. It has the objection of some additional weight to provide for its truck frame. Its motion is more favorable for the track, the coach and the passenger; and by its eight wheels may be, for a given number of passengers, lighter on the rail and more favorable for its durability; with bearing springs equal to those of the English coach, it would have superior ease of motion. It is, however, obvious from the general practice in this country, that these advantages have not been fully secured. There has been a taste for a large railway coach early inculcated and very generally perpetuated in our railway history. The prevalent idea seems to have been that a large coach gives importance to a railway, and little consideration has been given to the mechanical principles, or the true economy involved in this question. I think the most prominent claim of late for a large coach is that it is more safe; a position I regard as untenable. As before observed, the safety of a railway coach depends materially on its strength. A long coach body, having no interior cross supports, is necessarily more likely to crush in a collision than a short one having the same sized timbering. A coach body resting on its two ends is subject to the same laws of strength as a bridge; and it is fully appreciated that a bridge of long span, or length between supports, requires its scantling to be stronger in proportion to a short one, as the squares of their respective lengths; and hence a coach body 40 feet long between supports requires four times the strength to sustain the *same* weight that would be required for one of 20 feet between supports; and in this case, the strength must be further augmented to provide for a *double* weight, as the large coach is made double length for the purpose of carrying double the load. The influence of momentum has been noticed; it has been shown that the large coaches will feel its influence more severely than the small ones, and are more exposed to be broken up in a collision. No doubt, there is a size of coach that is most convenient for use—a size that best meets all the wants to be provided for. In this must be considered safety, cost, convenience of ingress and egress for passengers, facility of seats, handling at stations, and the traction of the coach. All these points can be better secured by a coach of 40 seats than one of 60 seats. I am well aware that railway men in this country will generally dissent from this view, for I have often discussed the question with many of them, and know the tenacity with which they adhere to large coaches, though I have yet to learn one single argument worthy of consideration that has been urged in their favor. The folly of hauling a house, with an audience-room capable of seating 60 people, through the country at the rate of 40 to 50 miles per hour, and calling it a vehicle for travelling, does not seem to have entered the general mind of railway managers.

Cars for freight have had a similar history as that described for coaches. The eight-wheeled, or truck principle, was adopted in this country for cars about the same time as for coaches; and has much the same advantages over the old four-wheeled railway wagon previously in use. The English adhere to the old single frame arrangement, and, I believe, generally with four wheels instead of six, as in their more modern coaches. They have spread the wheels longitudinally, from about four feet apart to six and eight feet apart. If it be, as it undoubtedly is, an advantage to reduce the weight on the wheel, as well as to secure other advantages, eight wheels are certainly better than four; and in regard to freight, not requiring high speed, it is difficult to see the reason that prevents English engineers from adopting the truck plan. As in coaches, we have generally adopted the truck principle in cars; but though we have, as I think, the best principle of construction, we have not made a wise improvement of it; for, instead of improving the opportunity of lessening the weight on the wheel, we have increased the load, so that no advantage in this respect is gained. Our cars are usually designed for ten tons of freight; and the car will weigh from seven to eight tons. On a few coal railways they have used eight-wheeled cars, that carry a load of three and a half to four tons, the weight of empty cars from one and a quarter to one and a half tons; but these are exceptions to the general practice, though they have been in successful use twenty years, and have resulted from superior engineering and management. The general tone of opinion among railway managers has been in favor of large cars as well as coaches. It seems to have been the prevailing idea, that the car should be as large as possible. The true question here involved is, undoubtedly, the car that will afford the means of cheapest transport. If the car furnish suitable accommodation to the freight carried, and is convenient for management, the essential requisites are provided for. The large cars are necessarily more severe on the track—causing greater derangement and wear than small ones; this must be obvious to the least consideration. The argument, claiming greater safety in large cars, is not tenable for coaches, and much less so for cars, not having even a plausible basis to rest on. No doubt there is a size of cars as well as coaches, best fitted for the service they are to perform. The car should be of sufficient size to permit convenient stowage of the freight generally transported. It is not incumbent, nor good economy, to provide, in the general plan, for those occasional pieces of freight that may be offered a few times in the course of a year; it should provide for the articles that constitute the great mass of the traffic. On the English railways, where there is as much necessity for this provision as here, they conduct all their traffic on cars ranging from twelve to fifteen feet long, and do not seem to feel any embarrassment. They transport live stock of every variety, and in large numbers; and the ease and facility of handling the cars at their stations, greatly impressed me when witnessing their operations.

If it be said, that it is at times necessary to transport long timber, or other article requiring a longer and larger car, let there be a suitable number adapted to this object, if it appear a branch of traffic that will warrant the expense. It would certainly be bad economy to incur extra expense on one hundred cars, that any one of them may be adapted to some peculiar form of freight requiring only the capacity of one. But there would not generally be one car-load of this kind of exceptional freight in a thousand, that would not be sufficiently provided for in cars well adapted to the general traffic.

Having been long convinced of the impolicy of large freight cars, some years since I proposed to stock a new railway with cars eighteen feet long for covered and twenty-one feet long for platforms; and contracts were made accordingly, and continued until about two hundred cars were placed upon the railway. The cars were satisfactory to the men who managed them—they were lighter in proportion to their useful load—more easily moved about the station by hand (of which more or less is always necessary), requiring less than half the force to handle them—no inconvenience was experienced in stowing freight—less labor was required in loading and unloading grain—they carried a greater percentage of useful load in the train, and in collisions were less injured than large cars, and the practical operation was satisfactory in every respect. But the directors of the company thought it would not do to have small cars, when other railways had large; alleging that it would give their railway a diminutive character, and injure the stock in the market; and, contrary to my advice, made contracts for as large cars as were in use on other railways. Against this, for a time, I remonstrated, urging that the small cars were working very satisfactorily, and that the value of the capital stock would depend on the net money that would be earned, rather than on the size of cars used, and that perseverance would show the true interest of the proprietors as affected by the size of cars. In reply, says one of the directors, "What do we care for the value of the stock twenty years hence?" I mention this, merely to show the short-sighted policy that often controls railway management.

Little as the subject regarding the weight of machinery has been considered, there is no one that more intimately concerns the economy of railway management. This cannot fail to impress the engineer who carefully examines the mechanical principles involved, and the result of railway practice. No superficial attention will prove a corrective; it must be thorough, careful, practical, and laborious, to reach a beneficial result. The wants of the traffic must be well studied; as they are to be provided for, and the traffic to be extended to new items, as fast as they may pay remunerative rates. This settled, and the important question follows, what sort of machinery will afford the means of doing it in the most economical manner?

In relation to coaches and cars, the question does not involve any difficulty when regarded in view of well known principles, and their proper adaptation to the use required. A coach of 40 seats is quite as convenient for making up a train as one for 60 seats. If a greater number are required for the train, they are as easily moved if in train; and if single, they may be moved more easily in correspondence to their capacity, and this is important so far as the movement is required by manual labor. The ingress and egress of passengers is in favor of the small coach; it may be lighter in proportion to the load carried, and it will often be sufficient for the fraction of a train; in which case it will serve all the purpose of a larger coach. Where, then, is the superiority of the large over the small coach which is to compensate for its more expensive maintenance? I have not been able to see it, though the large coach is adhered to with all the tenacity of "red tape." Some excuse it on account of the spring or elasticity of the long body, and claim that it rides more easily for the passenger than the shorter body. This is an admission that it is weaker than the short one, and, of course, so far less safe, as it could have no more spring if it was equally rigid. Now the body of a coach should be made so as to give it the requisite strength; and the springs should be under it, not in it. The truth is, the spring of the larger body arises from its weakness between supports, and the greater elasticity is not an object, but a mechanical necessity, arising from the desire to avoid the weight of scantling, that would be required to give it the same degree of strength that appertains to the more rigid frame of the short coach; and hence the resort to iron trussing in the body of the large coach to improve its stiffness, not its elasticity.

In regard to cars, there is nothing gained, even in convenience, in carrying from two to two and a half tons on a wheel, over such as may carry one and a quarter to one and a half tons, car and load included.

The real difficulty in this matter is the habit, that has become very general on railways, of using large cars and coaches. To remove this is no easy task, and considerable time must elapse before the ill-founded prejudice can be removed; and the engineer who may appreciate this subject will find it necessary to proceed cautiously with any changes he may desire to make; and to keep in mind that he will be opposed by men who have influence, although they do not comprehend the principles involved. From long observation and much reflection, I am satisfied of the substantial truth of the position I have taken, and that sooner or later it must come to be properly considered. Though I look appropriately to the professional efforts of civil engineers for improvement in this department, it must not be expected that all engineers will meet the requirements of the case. I have seen the regular daily operation of passenger trains on an important railway, under the advice of an engineer of extra profession of scientific claims, drawn by an engine of 20 tons weight, with two coaches nearly half filled with passengers, over a soft, unballasted prairie railway; the bouncing of the coach such as to alarm passengers not accustomed to this kind of travelling, and even to try the nerves of others. The discomfort of this sort of travelling is obvious, as the difficulty, or rather impossibility, in wet weather, of keeping the track in order, and the damage to the machinery, must correspond to its weight and the roughness of the track, and the net profits of this kind of management, if anything, must be very small. An engine of less than half the weight, and coaches of scarcely more than half the weight, would have performed the service perfectly, and so reduced the weight on the rail, that the expense of adjustment would be comparatively small, and the whole machinery kept in order at less than half the expense.

I have seen the operations of a railway that had no heavy business, either in freight or passengers, and the daily passenger train rarely containing more than one coach, and this drawn by a 20 ton engine. The little Saratoga engine, before described, weighing less than seven tons, would have taken in suitable coaches double the number of passengers, at the rate of 25 miles per hour, at half the expense incurred by this inconsiderate practice; and yet no one engaged in the management in either of the cases above mentioned seemed to suspect any want of sound economy. The practice arises, not from any inherent difficulty, but from the absence of careful thought, and from a ready yielding to careless and ill-advised habits.

On single track railways, there is an advantage on account of the inconvenience that may arise in a heavy traffic of too great a number of trains in having as much power in the engines as is compatible with a due regard to economy in other respects; and this will often lead to pressing the weight of engines beyond the best limit, and must be admitted to be some excuse for the practice. But this does not apply to the coaches and cars generally, for whatever be the amount of traffic supposable on a railway, the most favorable size of the coach will be essentially as good for a large as for a small traffic; and freight may be transported to any amount on a six-ton, eight-wheeled car, cheaper than it can on a larger one.

It is claimed by some that large cars are best for transporting live stock. Why are they better? Will the stock be more safe? The safest plan would be to have but one animal in a place; then there would be no other to injure it; and injury from each other is admitted to be the main source of damage, consequently the greater number together, the greater the risk. The English carry a large amount of every description of live stock in their 14 feet cars, and have not learned their inconvenience. I have not yet learned a valid reason for the imagined superiority of large over small cars for live stock, unless it be that when a car is once placed for receiving them, it would be more convenient to drive a whole train-load on, and save the trouble of changing cars in the process.

The constructing engineer should carefully consider the traffic he is to provide for, and aim, so far as he may have the power, at securing such machinery as will effect the most economical transport; and in this, I am well aware, he will find much prejudice. It is nevertheless highly important, and by degrees, the necessity of more strict economy will compel attention to this branch of railway management. On railways of large traffic, at good rates, the necessity will more slowly develop itself, but on those of light traffic, this attention to a more thorough scrutiny into expenses cannot long be delayed, and engineers should give this branch of supervision their most careful and diligent attention; bearing in mind, that it is not the largest machinery that should be run on a railway, but that which will afford the best net profit on the traffic they conduct.

CHAPTER XV.

OPERATING DEPARTMENT—INTRODUCTORY.

THE term operating has come to be understood as expressing the business of maintaining the railway and machinery and conducting the traffic.

In the general practice there is placed at the head of this department a superintendent, with subordinate agents to attend to the various branches of business. The subordinates are more or less varied, according to the magnitude of the traffic conducted. On prominent railways there are often assistant superintendents, who, under the general instructions of the superintendent, exercise his authority in his absence. On some railways a civil engineer is employed; but his branch of duty has not been, so far as I know, a very important one, in the operating department. There is a freight department, and a passenger department; an auditor and a cashier; or in some form their equivalents.

The superintendent directs in regard to all repairs of track and machinery, and for this purpose employs track-masters, and a master machinist; also constructs time-tables for the running of trains, employs conductors for running them, station agents, clerks, wood and water men, watchmen and switchmen. The auditor and cashier are not properly under the supervision of the superintendent, but receive their appointment from the board of Directors, and are responsible directly to the President or Treasurer.

The first business of operating a railway is to organize the conducting of the traffic, which must be immediately followed by the organization of a system for repairs and maintenance.

At the head of the freight department there is sometimes placed a general freight agent, but in others a chief clerk has the charge of the accounting duties under the direction of the auditor. If the former, it is usual for the general freight agent, under the direction of the superintendent, not only to attend to and supervise the freight traffic, but to make contracts for freight and perform other outside duties in this branch; in the latter case, these duties are more or less committed to the local agents at the stations, under the general direction of the superintendent.

The head of the passenger department is styled the general passenger agent. He keeps the accounts with all persons authorized to sell tickets, and receives the reports of the passenger conductors, also keeps the accounts with connecting railways in the sale of through tickets, which has become an extensive practice, and involves corresponding accounts.

CHAPTER XVI.

OPERATING—FREIGHT

On important railways, the freight department requires thorough system in the accounts. If strict attention is not given, so as to present full and explicit statements of the numerous parcels of goods received, forwarded, and delivered, confusion and loss will occur. The accounting duty of the freight department is by far the largest in railway operations, and demands the most careful and strict attention. In the early history of this business there was much confusion. Persons who had been employed in the ordinary methods of transportation by vessels and common roads, were often found quite inadequate to the rapid transaction required on a railway. One of the difficulties arose from the effort to make complete bills for lots that were distributed in several cars of the train. This has for some time been abandoned, and bills are made for each car, without reference to the particular lots to which they belong, at the same time aiming to keep lots together so far as may be. The goods for any particular station are usually placed by themselves in one or more cars, and such car or cars are left with their bills of lading at their destined station, to be unloaded and delivered by the station agent, who is held responsible for the charge and the proper delivery of the goods. If small lots of goods are to be sent to several stations, they are sent by a distributing car, and delivered to the respective stations from the train, which must stop a sufficient time to unload them at the stations. From the office of shipment a copy of the bill of lading is sent to the auditor, who charges the agent at the office of destination with the same. By this means the auditor's books show the accounts of the station agents, who are required to make daily reports to him of their cash receipts, and transmit the funds they collect to the cashier, or other fiscal officer. This makes the check complete; the auditor's accounts showing how much should be paid in by the agent, and the amount received by the cashier, and he is at all times able to show the Treasurer of the Company, or the board of Directors, the amount of funds that may be transmitted or paid over.

The station or other receiving agent may not collect the charges immediately, in which case his reports state the amount of goods on hand, and the charges due on the same. Occasionally, there will be lots of goods remaining for considerable time before called for by the consignee; and so far as not called for, the agent cannot collect the charges, and must report "goods on hand." If the agent improperly choose to keep the funds he has collected, he may do so by false returns as to goods on hand; and if suspicion of this arise, the auditor should send an agent to examine the goods on hand, and if he find the goods have been delivered he will know the agent has made false reports, and the correction may be promptly applied. With proper diligence on the part of the auditor no great loss can occur.

The conductor of a train has what is called a train list of the cars in his train, setting forth the number that are to be left at the several stations, and as he leaves the cars, takes a receipt for the same from the station agent, which is his voucher for their proper disposition; and if from any accident he is compelled to leave a car before it reaches its destination, he reports it at once to the office of the station where it was loaded and billed, and also to the general office, where means are taken to send it forward to its destination. In addition to the train list, the conductor usually carries the bills of the goods in his train, that are made to the station to which they are due, and these with the cars he delivers to the station agent. In some instances, the bills are sent forward by a passenger train, and received by the agent before the goods arrive: in some instances it has been the practice to send forward the train before the bills were made out; in such case the bills follow, and may not arrive until after the goods reach their destination. It is a bad practice and leads to embarrassment in the delivery of goods, and should not be tolerated.

The freight business is now so well organized, that losses seldom occur from stray goods, and the mode of checking appears to be complete.

CHAPTER XVII.

OPERATING—PASSENGERS.

In the passenger department, the accounts are kept by the general passenger agent, who supplies from his office tickets to all persons authorized to sell them, and the checks in this department, so far as relates to ticket agents, are complete. Deception can hardly be practised except by collusion with a third party.

The conductors of the passenger train report to the general passenger agent, and there has not been found any means of checking him for the collections he makes in the coaches, except so far as it has and may be done by espionage. In consequence of this difficulty most railway companies offer inducements to passengers to procure tickets at the offices by allowing a discount of 5 to 10 cents on a ticket. This induces a largely increased purchase of the ticket agents; but there is a considerable amount still collected in the coaches by the conductor, which is consequently dependent on the integrity of his return. I know of no means to wholly remove the difficulty arising from this want of a check on the conductors.

On English railways most of the stations are inclosed, and the passengers on leaving and usually on entering the station, must in the former case give up, and in the latter show their ticket; but there is nothing to prevent the ticket collector from taking the fare, and if the passenger has no ticket, the integrity of the collector must be relied on, to account for the money. In large towns a different course is pursued: about one mile from the station, a ticket collector gets on the train and collects the tickets, the train being delayed time enough for this purpose. Here the collector occasionally finds a passenger without a ticket, and I have seen a collector in such a case collect the fare for the distance given by the passenger, without remark, and have concluded it not to be a rare occurrence. Of course the return of such a collector can have no more check than that of a conductor. The English are very watchful of the entry of passengers into the coaches; but in the large towns I have seen no great difficulty in a passenger getting into the coach of a making-up train; the passage from the ticket-office to the coach shed being open and often without a doorkeeper, affording no impediment to his entering the coach, and if asked when in the coach if he had a ticket, was seldom required to show it, the man engaged in seating the passengers usually being content with the word of the passenger. I have known instances in which passengers had no tickets, but paid fare to the collector at the large town station. Though the English system does not appear to be complete, I think a much greater proportion of fare is paid at the offices than on our railways. If the English system required, in the case of a passenger not having a ticket, that he should go to the ticket-office and obtain one before he could pass the gate either out or in, and not allow either gatekeeper or collector to take fare in any case, it would seem to render the check complete. This would require the same practice at the terminal and other large stations, as at the smaller ones. I have supposed the practice of sending a collector before entering the large towns, was introduced to avoid delay to the passengers, who take carriages in the station-yard; and so long as it prevails, the system of checks must be imperfect. If the English system was carried out as above suggested, it would not allow a fare paid, except to a ticket agent, and the collector would be merely a collector of tickets. Some effort has been made in this country to pass the entering passengers through a gate or door, and there requiring them to show their tickets before they pass into the station, and I have often gone on a train directly from such a station, and seen the conductor quite busy in receiving fare from those who had no tickets. The fact is, it often happens that a crowd is collected around the doorkeeper, and passengers have some delay in finding their tickets, and with sundry items of hand baggage, and sometimes (often) infants, and children but little more advanced, causes delay and impatience, offering inducement and opportunity for others to crowd the passage, and more or less pass without showing, and without even having tickets to show.

I see no effective mode of forming a check on the passenger receipts, without complete inclosures, so arranged that no person can go into the car-shed or yard of the station ground, without passing a door or gate arranged with one or more openings, and a tender at each, so that no greater number need pass any tender than he could examine and see that they had tickets; for most stations one passage would be sufficient; and the arrangement would require to be extended to all stations where passengers were received. There would still be one source of evasion; namely, a passenger may procure a ticket for the next station, and continue on to a more distant station. The conductor would, of course, detect him; but how is the fare to be collected beyond the station ticketed? If the conductor collect, as is the present custom, then the check on this fare is lost. To carry out the system, the passenger should be treated as one that refused to pay his fare, when he extended his passage beyond the station ticketed, and put off the train, unless he could satisfactorily explain to the conductor the fact of not possessing a ticket, and would agree to procure a ticket at the next station. But who is to control the conductor in such a case, if he choose to take the fare, and allow the passenger to proceed on the train? As the conductor is the only man of authority on the train, from the beginning to the end of his route, no direct check can be had on his proceedings with a passenger on the way, although there would be an indirect check in the fact becoming generally known that the conductor was not in any case allowed to receive fare. The fact of receiving it would thus be likely to attract the notice of passengers, and hazard a report that would bring the conductor to an account. This might not, in all cases, be a protection, but it would be a salutary check.

If, upon any plan, the exit gate be used as in England, to insure the full collection of fare, the tender should not be allowed to take fare from any delinquent passenger; his business should be confined to the collection of checks or tickets, and if that is wanting, the passenger should be required to procure it at the ticket-office. The exit gate is objectionable at large stations, and hence the English railways only adopt it at the secondary stations, and depend on a collector at the large stations, as before explained; and on the collector there is no check, any more than on a conductor.

It must be conceded that the crowd and hurry that occur at important stations, in making up trains, and more especially at the breaking up of the train at terminal stations, involve a difficulty in any system that can be relied on as a perfect check. The entrance gate is the most easily guarded, and if this part were well arranged, there would be but a small exposure to loss. It involves the necessity of receiving the baggage at or near the ticket-office, in order to save the passengers from delay and confusion, in attending to both ticket and baggage checks. It is often the practice to allow friends to pass the gate with passengers, with a view to see them seated in the coaches; this is obviously an error, and should not be allowed on account of its liability to abuse. With proper arrangement of inclosure and gates, very nearly all the fare will be collected at the ticket-offices; and if the conductor and all collectors of tickets were prohibited, in all cases, from taking fare, making it the duty of the former to require any delinquent passenger to go to the office of the next station and procure his ticket; or failing to satisfy him that he had a fair excuse for his delinquency, to put him off the train, as is now done if payment is refused, would leave but a small deficiency in checking the passenger receipts. No system can be safe from collusion; but this is a far more difficult practice under a plan like this proposed, and with proper care in the selection of agents, no great or material delinquency will be likely to occur.

In all business involving trusts, the importance of a system of complete checks in reports and accounts is very obvious; to both parties it is beneficial, securing to the principal full accounts of his dues, and to the honest agent the means of showing the fidelity of his proceedings. By the existing practice the faithful conductor has no power to prove his fidelity, nor the railway managers any exact means to prove the truth of any suspicions they may entertain, except by espionage, which is not desirable if it can be avoided—a state of things often very embarrassing to both parties; and action, when taken, is usually based on conjecture, and in many cases no certainty can be arrived at. The reports of conductors may appear fair, but as there can be no proper check to verify their accuracy, they afford no conclusive evidence in the case either way, and the unfaithfulness that may be supposed to exist, generally rests on suspicion, arising from general indications, as habits of dissipation and expenditure, or the reputation of acquiring property faster than is compatible with the salary received; and these facts may or may not be known, according to the prudence of the agent in keeping them out of the sight of those who are interested to know them. It must be the desire of all honest conductors, that their business should be placed on a system of the most perfect checks; and the railway company, in order to protect themselves against such as may be unfaithful, and be able to know those that are trusty, should adopt the most efficient system practicable, for securing the fidelity of their passenger, as well as their freight receipts. There is, no doubt, a difficulty in attaining this end, and it will involve considerable, though not serious, expense to prepare the stations properly, so as to meet the requirements of any efficient system. With faithful men for conductors, no great evil will arise from a partial collection in the coaches; but, as before observed, it is a difficult, if not an impracticable thing, to discriminate between those that are, and those that are not faithful; and therefore the temptation, as much as possible, should be removed.

In addition to the fidelity of his fiscal duties, the conductor has important duties in the management of his train. His good manners and attention to the incidental wants of passengers will have much to do in promoting their comfort and convenience, especially with the infirm, and with females who may be without an escort. He should be a man of good sense and ready sagacity, in discovering how he may be useful in these respects, as upon his judgment and promptness the reputation of the railway with the travelling public will a good deal depend. Upon him rests mainly the responsibility of a proper regulation of the stops of the train; to see that due notice is given to the passengers that they may be ready when the train stops to leave with the least delay, and that passengers to be received are promptly got on, so that there be no unnecessary delay of the train; for what is lost here must be made up by increased speed between stations, which it is important to avoid. The system of marking, registering, and checking baggage is now so well arranged that the baggage-man may have the whole assorted, and that for the next station ready to be delivered the moment the train stops; and as that to be received is prepared by the station agent, ready to be handed in as soon as the baggage-man on the train has discharged that to be left, very little time is necessary for baggage, and usually it can be discharged and received on board by the time the passengers are changed. If the train is not delayed to take on fuel and water, the stop need not consume over a half to one minute, according to the traffic of the station. This may seem a small matter, but the character of the management of a railway may be determined by the manner their stops are conducted. The baggage-man and brakemen are under the supervision of the conductor; and much depends on the promptness and judgment of the latter in stopping the train; at all times they should be quick on the signal, and exercise discretion in stopping the train as soon as possible, at the same time, not to cause the wheels to slide, except in cases of danger to the train. The engine driver is also under the supervision of the conductor to a certain extent, to see that his duty is performed skillfully in stopping and starting his engine—in maintaining uniformity of speed—in entering or leaving curve lines, so as to avoid disagreeable lateral motion to the train. These things are not only important to the easy motion of the train, but to the saving of repairs in machinery and track, which are unnecessarily injured by the jerking motion and irregular speed caused by an unskillful running of engines. The conductor is the captain of the train, and if any of the men employed on it fail to obey his directions, or transgress the regulations of the managers, it is his duty to report the delinquent to the superintendent.

From the preceding remarks, it is obvious that the conductor should be a man combining fidelity with courteous manners, general intelligence, and sound practical sense; careful to avoid danger, and skillful to extricate his train in case of difficulty; he must be unknown at saloons of dissipation—an indulgence which would inevitably impair his character for fidelity to his trust, and should be carefully avoided.

CHAPTER XVIII.

OPERATING—ENGINE DRIVERS.

Competent engine drivers are indispensable to good railway management. A man may be a good mechanic, and understand the machinery of the engine well, and still be a poor driver. It is necessary that he have good judgment as to time and speed: this is less important in running very high than in running low speed, as the former usually requires the full power of the engine; but at a lower rate of speed, he will be in danger of arriving too early or too late at the stations, or will run too rapid a part of the time, and too slow at others; and as it is the interest of the railway to run the lowest speed that will answer the demands of the traffic, all unnecessary high running is a damage to the proprietors; and it is therefore an object to obtain the greatest regularity in speed. With a skillful driver, jerks are rarely felt in starting—no running past the station where a stop is to be made, nor more breaking up than is necessary, especially such as to slide the wheels, which is very severe on both wheels and rails; coaches and cars are injured by sudden starting and stopping, and great care should be exercised to guard against these evils as much as possible. In passing to and from curves in the line of track, the skill of the engine driver is to be exercised, and will demonstrate how well he understands the handling of his engine: if well done, the motion will hardly be noticed, and if ill done, a disagreeable lateral motion will result, as may often be noticed. This is a test that few drivers can stand well. It requires a good eye on approaching or leaving a curve, and power and judgment in handling the engine. I never notice this duty well performed without a desire to express my satisfaction to the driver, as I regard it as a high exhibition of skill in this department.

On express passenger trains the work of the engine driver is of short duration, and not over one or two operations of taking on fuel and water will occur on his route. But on way-passenger and freight trains, stops are more numerous, and much more time is taken to perform the route. It is of great importance that the business at these stops be dispatched in the least time practicable; for, as before observed, whatever is lost in time here, must be regained by so much increase of speed. In this matter much will depend on the conductor, but the usefulness and value of a prompt and competent engine driver is very obvious; his fuel and water should be taken on, and his engine oiled with the least delay, so that he may be ready to move promptly on the signal. A freight train at the station will often have more or less cars to be taken out, and others put in the train, all of which should be done with the utmost promptness, to save all possible time for appropriation to moving on the railway. It should be the constant care of the managers to inculcate on the engine drivers the importance of occupying as much as possible their card time in movement on the track, and to establish a judicious discrimination in favor of those who most effectually discharge their duties.

The interests of railway proprietors are materially affected by the skill and fidelity with which the engines are run, and the very general practice of paying the drivers all alike is incompatible with sound management. There is an important difference between a first-class, and a fair engine driver, and of course a greater difference between a first-class and an ordinary driver: and though this seems to have received very partial attention, there can be no doubt, that a judicious system of grades, under the supervision of a competent foreman or master mechanic, would produce a salutary and laudable spirit of emulation—would give satisfaction to the best and most competent, and show the careless and unskillful, the necessity of effort, cultivation, and fidelity to reach the first rank. A knowledge of the manipulations required to start and stop, and to perform the current operations in the ordinary work of running an engine, is but a small part of the accomplishment of a first-class engine driver. He must know how to feel his engine, and apply the power judiciously to the work required; and the skill to do this well, is only acquired by experience and good judgment. The objection has been raised, that it would not do to discriminate, and that the attempt to establish grades of compensation and service, would produce dissatisfaction; but clearly not with the first class; and I can see no force in this objection, other than appertains to any branch of business. Certainly, a duty that involves so much of fidelity and capacity as this, should not be left to an indiscriminate equality in rank and pay, treating the best with scarcely a consideration above the least competent; a rule that would be discarded as inefficient in any other branch of responsible business. To do any business well, requires attention, firmness, and sound discretion, and that is all that is required in this.

CHAPTER XIX.

OPERATING—REPAIRS OF TRACK.

The repairs of track will require attention as soon as the running of trains commences. If the railway has been thoroughly constructed, these will be very light at first, and the running may continue for a considerable time before the expense is great, and will be confined mostly to the adjustment of rails, keeping side drains clear, and fences in order. The bridges, even if of timber, if they have been well constructed, will require but little attention for some time. According to the weight of traffic, the expenses will gradually and pretty steadily increase. It however rarely happens, that our railways are opened for use, on a thoroughly constructed work. The more general practice has been to commence the running as soon as a train may pass with tolerable safety—the track imperfectly adjusted, with little or no ballast to support it—the side ditches imperfectly opened—the approaches to bridges indifferently secured, and station buildings more or less unfinished, even if commenced. It is therefore obvious that the current expenses of maintenance and repairs will be, at the outset of operating a railway like this, much greater than on a thoroughly constructed work, without regard to those expenses that are properly chargeable to construction.

The usual mode of conducting repairs of track, is by track-masters, acting under the general instructions of the superintendent, or his assistants. The length of railway assigned to the care of a track-master, varies from twenty-five to seventy-five miles, according to the condition of the work and the judgment of the superintendent. It is usual for the track-master to divide his forces into parties of four to ten men, with a foreman over each, occasionally bringing several of these together to man a gravel train, or for other work requiring a large force. He will have one or more carpenters to look after and repair bridges, and these are sometimes required to attend to repairs of buildings at small stations. Tools and materials for the track-master are supplied by the storekeeper of the company, who keeps an account of what he furnishes, charging the same to the track-master. Such articles as lumber, cross-sleepers, etc., may be most advantageously obtained along the line of railway, and the track-masters frequently purchase under the direction of the superintendent.

It is the duty of the track-master to understand the condition of the track, and all the structures required for its support: to see that his men are properly distributed and arranged for efficient work, and that the foremen are capable and reliable men. The latter becomes more important as the length of his district is increased, and his time for personal supervision over each party is reduced; as in this case he must depend more on his foremen than would be necessary with a less extent of district. It is not always the practice to place the repairs of buildings at way-stations in charge of the track-master—this is sometimes placed under the care of a general superintendent of all the buildings on the railway. The track-master is sometimes charged with the purchase, inspection, and measurement of wood for the engines. It is his duty to certify the monthly check-rolls of his men, and all bills of wood for fuel and articles he may purchase or inspect; these are also certified by the superintendent; but it is evident he must rely mainly on the track-master for the accuracy of the check-rolls and bills, though an active and vigilant superintendent may detect important errors, and exercise a salutary supervision.

In addition to the duties above stated, it is the duty of the track-master to see that signals are placed at any points of danger that may require the train to move with caution or come to a stand; and if a train be thrown from the track, or break down, to promptly supply sufficient force to clear the track and put affairs in order. Much depends on the fidelity, vigilance, and good judgment of this agent; he has large expenses in his department, and his forces are necessarily scattered over a large district. There has been much diversity of opinion as to the proper length of district for a track-master. This doubtless depends measurably on the nature and weight of traffic, the character of the railway, and his own capacity and efficiency for business. If the district is long, he will take the opportunity of trains to pass from one point to another, and seldom see his track, except at points, only as he sees it from the train as he rapidly passes over it. This practice may properly be adopted in occasional instances; but where it is the means used for the greater part of his movements for the inspection of his track and the supervision of his men, it necessarily affords him but an imperfect knowledge of its condition, and renders him more dependent on his foremen, a class of men not expected to possess the same capacity and judgment as himself, and therefore practically lowering the intelligence it is designed to give to the supervision of the track. There is but one way of movement over his district that will enable a track-master to properly inspect his track and superintend his men—namely, on foot; even the hand car should not supersede walking as the usual practice. The hand car or the train may occasionally be very proper, when it is necessary to go quickly from one point to another; but at least, as often as once or twice a week, the track-master should pass carefully on foot over his district, inspecting carefully the condition of the track and other works, and be able to consider well its condition, and give full instructions, devoting as much time as may be necessary or practicable to his men, in order to understand well what has been done and what is necessary to be done. There can be no doubt, the want of thorough supervision in this department has often been the cause of very serious accidents, causing much expense that proper attention would have saved. The mere neglect of a side fence has opened the passage for an animal to get on the track and throw off the train. This is especially dangerous in the night, when animals are prone to lie down between the rails, assuming the most formidable position to throw off the train with disastrous violence.

One of the first things that indicates an inefficient track-master is the neglect of the side ditches, and consequent imperfect drainage of the road-bed. This neglect will not fail to increase the expense of keeping the track in order, and at times of rain, render it impossible to maintain it in good adjustment. To remedy this, more ballast must be provided to give temporary support, often at heavy expense, and ballast cannot give a good track until the draining is properly attended to. A light ballast, with good drainage, is far better than heavy ballast, surcharged with water.

It often happens, that in the construction of a railway, even if the general plan is of a thorough order of work, that circumstances lead to the adoption more or less of temporary structures, that will soon require the substitution of more durable, if not more substantial materials. It may be that building stone could not be had in reasonable proximity to the work, and timber had to be used for culverts and bridges, and sometimes for the foundations of buildings. As noticed under the head of "Constructions," there is no material equal to durable stone for culverts and bridges; and when it becomes necessary, for want of this material, to use timber, it should be regarded only as temporary. So far as timber may have been adopted, the process of superseding it with more durable materials should be commenced as soon as those materials can be commanded, and the work proceeded with as circumstances may require. It may be assumed that substantial works of timber may be relied on from five to ten years, according to durability and exposure; and by commencing soon after the railway is put in operation, the work of substitution may be carried forward gradually and with economy. The engineer of the company should carefully examine and decide on the most important structures for the commencement of this work, keeping in view the current necessities of the track, and the probable comparative durability of the structures. It may happen that the finances of the company will not admit of a complete substitution of all during the life of the timber structures; in which case the engineer should proceed on those most important, and so far as renewal with timber may be necessary, see that this is done in good season, to effectually guard against any giving away that might cause accident to the passing trains. Wooden bridges are often built over valleys that have very small, or perhaps no permanent water-courses, and are proper situations for culverts of stone surmounted with an embankment of earth. I have known cases where the bridge cost nearly, and sometimes quite as much as the culvert and embankment would have cost, if stone could have been had for the culvert in the vicinity of the work. In considering the renewal, the question will probably be entirely changed by the facility the railway will afford in transporting stone from distant quarries, which by ordinary roads were out of reach in the original construction; and the economy of the case in renewal would be entirely changed, requiring thorough examination, so that perishable works be not rebuilt in situations of this kind, where no material saving in cost can be secured by adopting the original plan. In no case should a wooden bridge be neglected, if a doubt may exist as to its safety for the passing trains.

As the work of renewal with durable materials goes on, the railway will be steadily assuming a more substantial and permanent character, rendering it more secure for the passage of trains, and diminishing the expense of repairs and maintenance.

In the maintenance of track, a very important duty should rest on the engineer, to see that the original drainage proves adequate for maintaining a dry road-bed; and if not, to devise and proceed with measures that will effectually secure this object; a branch of railway affairs that demands the most diligent attention. As this, together with the kind of material for bridges, has been pretty fully discussed under the head of construction, it does not appear necessary to say more in reference to them in this place. Whatever is necessary for the construction of a track and its appurtenances, is to be maintained in the work of repairs and renewal.

For a time after a railway track is opened for use, no great expense for the renewal of rails and cross sleepers will be required. The length of this time will depend on the quality of the materials, the amount of the traffic, and the weight of machinery used. As decay and wear produce their effects on those materials, the expenses of repairs and maintenance must necessarily increase, until they reach a point that will require an annual expense, varying each year, but forming an approximate average in a series of five years. The sleepers will begin to fail ordinarily in about five years, while a portion may last eight years; the rail will be very dependent in its wear on the weight and speed of the machinery used; the latter will be more important than the extent of the traffic. The circumstance of increasing expenses, at least for a few years, indicates the propriety of holding in reserve a portion of the net earnings, to meet future expenses, if it be the purpose to maintain regularity and uniformity in dividends to the proprietors.

CHAPTER XX.

OPERATING—REPAIRS OF MACHINERY.

The English term is "rolling stock," frequently termed "the plant." In this country, the terms "equipment" and "machinery" are both used. I see no more propriety in applying the term equipment to railway machinery than to cotton or woollen machinery; it is purely a technical term in military affairs, and its application to civil, is an innovation that may be applied with as much propriety to all other as to railway machinery; and it would be much more appropriate to apply it to farm tools than to the machinery of a railway. Rolling stock is a more expressive and better term; but I consider it fully entitled to the term "machinery," and therefore use it. There is a portion of stationary machinery used in the railway repair shops, but what is understood generally by the term railway machinery is that adapted to motion on the track.

The machinery of a railway peculiarly requires to be kept in good order; and as soon as the operating service is commenced, this kind of repair will demand attention. With a new stock in good order, it should not for some time involve heavy expenses; but they will steadily increase, and eventually become a large item. Unless when radical changes are found expedient in the plan of machinery, the repairs will be a perpetuation of each machine. There may, and probably will be cases where a coach or car may be so entirely destroyed as to leave nothing to rebuild upon; but I have never known an engine to be so destroyed that there was not sufficient left on which to rebuild. In the ordinary deterioration, it will be found that new parts may be advantageously ingrafted with those that have suffered a comparatively moderate degree of wear, and still sufficient for much useful service. In this way one part after another is substituted, until there may be nothing of the original structure left; still it is practically the same engine, coach or car, with the same name or number; the repairs, therefore, involve the renewal and complete maintenance of the machinery, and will demand shop accommodations and stationary power sufficient for this object.

This department of business is placed under the supervision of a master-machinist, who is not only charged with the works of the shop, but with the direction of the engine drivers and firemen, as to their order of service and their general duties. It is for him to judge of their capacity and grade of service, and to correct any delinquency in duty. On some important railways there is a master-carpenter, who has charge of all woodwork, and is independent of the master-mechanic; on others, the master-carpenter is subject to the general supervision of the master-machinist, though the latter has no specific charge of the workmen in the carpenter's department. The latter plan I regard as the best, as it is better to have a single head in such an establishment, where the works of the two branches must be brought together, and the order of work in each should be so directed that delay in the work of one may not embarrass the progress of the other, or disturb the efficient harmony of the system of which this department is a portion. This leaves the master-carpenter independent, so far as relates to the employment and control of his men, and in the management of the affairs of his shop, subject only to general direction in the order and plan of work as the master-machinist may regard proper.

The master-machinist has the general supervision of such secondary shops as, in some cases, are established at the end of routes for the purpose of making small repairs, required to keep the machinery in use, or to enable it to run to the principal shop.

The master-machinist requires a clerk and time-keeper (those at shops of small business may be both in the same person), to enable him to keep his accounts so as to charge to each machine the expenses incurred on its repairs, and for the general accounts of his department. It is indispensable to a good administration of business, that a strict account should be kept of the cost of repairs on each machine; this will show the quality and value of each—and is particularly important in respect to engines, as it will indicate the care and skill with which they are run. It will also show the expense of each department of traffic, so far as respects the machinery, an item of information important to be known by the superintendent, that rates may be properly adjusted to the expenses, and that he may know to what extent he may go in cultivating a doubtful branch of business. On this plan the repairs of passenger machinery and that for freight are kept separate; a distinction that cannot with exactness be maintained in the repairs of track, though quite practicable for machinery, and should not be neglected.

The master-machinist will require a number of foremen, who should be experienced mechanics, to take the special charge of the different branches of work.

In addition to the work of repairs, it is the duty

of the master-machinist to keep an account of the miles run by each engine—the oil and waste used by each of the engine-men; and to these is, in rare cases added, what should always be done, the amount of fuel used by each, which will be further discussed hereafter.

The master-machinist occupies a station of much responsibility; to fill it properly, it is not merely necessary that he should be a good mechanic; he should add to this a knowledge of the essential principles of machinery, and a good general knowledge of business, that he may comprehend the system required to carry forward with order, regularity, and efficiency the various works committed to his charge. He should also be able to judge of the qualifications of his workmen, and to enforce proper discipline; the latter is peculiarly important in respect to his engine drivers, who, during most of their time on the trains, will be out of the range of his personal inspection; and on whose fidelity and skill much will depend in the repairs of machinery. To a railway of any considerable importance, it is poor economy to employ an inferior man as master-machinist, on the competent and faithful performance of whose duties so much of the economy and efficiency of the operating department depend.

CHAPTER XXI.

OPERATING—CIVIL ENGINEER.

There are some railways that regularly employ a civil engineer, and others that do not. In some instances the superintendent of the railway is a civil engineer; but his duties are necessarily so general, that he can give but little attention to those peculiar to engineering. Some managers are prejudiced against the employment of an engineer after the railway is put in operation, which is a short-sighted view of the subject, and not at all consistent with a proper knowledge of the duties required. It is true, some railways are managed without the aid of an engineer; in such cases various works will be put up by ordinary mechanics, who sometimes manifest very good judgment, more particularly on works of minor importance; but in general (not that the mechanic cannot execute well any work in his trade, but from want of the experience required, he may not comprehend the kind of structure most suitable for the object), it will be found that they are deficient in system, often entail much unnecessary expense, and fail to produce the regularity, economy, and efficiency that would result from the supervision of a competent engineer. It is the engineer's business to study thoroughly the wants of the railway, in regard to the track and its appurtenances—the effect of the action of the machinery upon it—the kind of machinery that will produce the most economical transportation—and the most favorable arrangement for the stations, station-buildings, and shop accommodations. He should be the authorized adviser of the master-machinist and track-master, and superintend all contract work for renewals of bridges, culverts, buildings, and machinery. By his connection with the track and machinery, he will be able to judge of the effect of one on the other, and by general observation and carefully conducted experiments, will come to a more thorough understanding of the interests of the railway, than will be likely to be reached in any other way.

These views may fairly be urged on general principles; as a man who is educated to a particular business—whose time is devoted to a full understanding of its requirements—and who is stimulated by the consideration of professional reputation, is more likely to conduct affairs advantageously than one who picks up his ideas at random, and though doing some things very well, will probably often fail in respect to others. Certainly, the important matter of maintaining the track and machinery of a railway should be committed to the most competent hands. My own experience and observation fully sustain the propriety of the course of management recommended. Any other course is precisely like giving the preference in a business to a man who has not made it his study, or practised the arts involved in its current transactions; a course that is not practised by intelligent men in other departments of labor or business. To this argument it may be said, that men have managed railways successfully who have had no previous training. The reply to this is: That such have mostly been on short railways, with little or no rival interests that called for close and strict management; and that they have depended on the master-machinist for all that related to machinery, and on the track-master for the management of the track; and though by a ready facility of adaptation they have succeeded very well, there can be no doubt that they managed better as they acquired knowledge and experience. The success of such may be compared to that of a man, who, never having learned the trade, begins to make shoes. At first he makes slow work, getting information any way he may, and by perseverance makes progress according to his aptness for the work, and though he spoil much leather, eventually may become an expert shoemaker, having lost much time and material in acquiring the art. So it may readily be perceived, that the company must expend more or less to educate their man, if he has not previously been educated. The evidence of this I have often seen in the machinery used on railways; very powerful it may be, but so damaging to the track and the repair shop, that if it could not be appropriated for its material, the interest of the proprietors would be best consulted by dropping it in some abyss, too deep to admit the hope of recovery.

The perfection of machinery has thus far been dependent mostly on the machinist, and on the competition of rival manufacturers; and in this way much has been done to make powerful and effective machines. I have known some very intelligent and excellent machinists who have contributed largely to improvement in the arrangement and workmanship in this department, to whom the public and the railway interest are greatly indebted. At the same time, I have rarely noticed in them any special concern as to the effect of their machines on the track.

It has been the general practice to commit the care of machinery to the master-machinist, and this is proper in all that relates to manipulations, and, to a large extent, to the plans of work; but, as before observed, this class of men rarely give much attention to the influence of their machinery on the track, nor can this be expected, as they have no supervision of the track. They regard the power of their machinery as the best and proper indication of their ability as machinists, and consider that the track should be able to bear it; and if it does not actually break down under the service of the train, the machinery is regarded as all right—draws large trains and runs high speed. The machinist has no care, and takes no note of those every-day expenses that are required to keep the track in adjustment. Then the track-master has no charge of the machinery, and rarely realizes that it involves any question in relation to his duties, but goes on as best he may to make his track capable of sustaining the service. Thus nothing can be more clear than that the track and machinery should be under the general supervision of the same man, who should be capable of comprehending not only the adaptation of one to the other, but the service of each in effecting the most economical transportation of the traffic to be provided for. To merely run trains, is a thing that may be done with small experience: To run trains and manage the track and machinery, so as to effect the most economical transport, is a very different thing, and as yet very imperfectly studied.

I am well aware, that a large proportion of railway superintendents will not concur in the views here advanced. For some reason which it is not necessary to discuss, they seem to have an aversion to civil engineers, and usually contrive to get them off the railway, or restrict their authority to a very limited field. They seem willing to see them employed in hunting up maps and examining doubtful questions on right of way—occasionally to set levels, stake out work, and compute contracts; all of which are their proper duties; but further than this they will not tolerate. Thus, under these circumstances, the engineer cannot enter on the main and more important field of duties which his profession should enable him to fill more effectually and beneficially than any other person.

In the duty of selecting an engineer for the responsible charge here recommended, it is necessary to exercise the same scrutiny, the same practical sagacity, that is called for in other departments of business. What, then, is the course ordinarily pursued by business men? If a man has on hand an important litigation, he does not take the first lawyer he meets, nor in case of ill health, the first physician that may chance to be named, or come in his way. He is not satisfied with the mere professional title. The title of lawyer or physician is good as far as it goes, or is sustained, and the same for an engineer. But the man sought is one well skilled in his profession, and capable of rendering the special service for which he is wanted. A business may be badly done by a professional man, and the responsibility of failure may rest mainly on the employing party, who has failed to exercise due discretion in the selection of his agent. Men may, and often do, bear the professional title of civil engineer, as well as of lawyer and physician, with very slender qualifications. It is not the mere scientific engineer, who may bewilder with hair-spun and useless calculations, nor the practical engineer, who may be able to collect statistics, run levels, set pegs and stakes, copy drawings, and make out estimates for contractors, if these be the end of his accomplishments, that is needed. For the object here proposed, the engineer should be familiar with mechanical principles—understand well the strength, durability, and adaptation of materials—by experience, observation and study, should have a fund of practical information at command, that may be available as occasion or emergency may require. It is, moreover, necessary for him to be a good business man, familiar with the varied interests of a railway, in which he will find ample scope for sound practical sense and experience in his dealings with both men and things. Not a man that is punctilious of his dignity, and unwilling to profit by a good hint from the most humble workman, but ready to rest his standing on his good sense—his frankness and uprightness in intercourse with others. Such a man will never have occasion to complain of a want of respect in his business intercourse. The list of criterions would be incomplete if the too-prevalent vice of a speculative spirit (the quicksand that has sometimes destroyed or perverted the best talents of an engineer) was omitted. An engineer should be so completely content with his salary that his mind would be wholly devoted to the work placed under his charge, and his own happiness as much concerned as that of the proprietors in obtaining a favorable result for the enterprise. Not less than for a lawyer, physician, or merchant, an engineer, to succeed well, must enjoy his profession, and find his chief recreation in the cares, duties, and results of his labors.

It will be asked: "Where are you to find men of this grade for all the railways?" Certainly we do not find all, in any profession, to come up to this standard, and probably no more can be said for that of engineers than for others; and the same course must be taken in one case as in others. A man will seek an agent according to the importance of the service he requires; one of moderate or high attainments, as in his judgment may be necessary for the business he has to provide for. There is no difference in providing for a railway; the most important will demand the best experience and qualifications. First of all, see that the engineer is a man of prudent habits, sound practical sense and fidelity—such a man, with fair elementary qualifications and experience, will fill most of the duties required, and steadily improve by experience in the special service that may be involved in the work required to be managed; and although he may not at first be able to fully meet the requirements of the situation, will, from his education, have a ready adaptation, and be likely to improve in qualifications, so as to meet all the duties and successfully conduct the business that may be committed so his charge. No engineer can go upon a new work and not find something peculiar, that will demand his careful reflection, and the deliberate consideration of any advice that he may receive; and nothing so fully reveals his incapacity as a pretentious assumption of knowledge, claiming to understand everything. In short, the same business sagacity that is required in other important affairs is required in this, and will have the same prospect of being rewarded with success.

I have been thus particular on this point, from having observed very indiscreet selections, and from the conviction that incompetent engineering should be held in the same light as incompetency in any other profession; and from the conviction I have long felt, that much of the prejudice entertained against their proper position in the management of railways, has resulted from or been perpetuated by injudicious selections.

The difficulties in railway management, noticed in a former part of this work, arising from their recent introduction and rapid extension, has necessarily led to much imperfection, and the engineering profession has not kept pace with demands so hastily urged upon its attention; and having been mostly occupied in works of construction, engineers have not generally had time and opportunity for that close attention that is necessary to render them as useful as they should be, in the maintenance of railways. Notwithstanding this deficiency must be admitted as of general application, it does not change the position of this question; for when they are placed in their proper position in the management, they are, from their professional training, experience, and habits of careful study, more likely to succeed in perfecting this branch of service than men who are not especially fitted by professional acquirement.

It will be admitted that training of some sort is necessary for every department of labor or business. It must also be admitted, that youth is the most favorable, as well as the most economical season for acquiring elementary knowledge. At a later period in life, the mind is diverted by various objects—time is more valuable, and elementary learning, if it be sought in the age of maturer manhood, will be sought as a necessity, not as an appropriate pursuit, and is likely to be hurried and imperfect. The same may be said of any branch of art; there is no time of life for the training in elementary art or knowledge so valuable or economical as youth. Though there are exceptions, it is a general truth, that it is expensive to learn a new occupation in maturer manhood; and, consequently, the railway company that commit their business to unskilled or uneducated men, must be at the expense of educating them during their supervision of business, and while they are learning its arts and duties; meanwhile depending on advice, trusting to the guide of others as they may chance, to find out matters beyond their own powers of criticism. Can there be a doubt that the proceedings of the pupil will often be undecided, wavering, and wanting in that system indispensable to the efficient and successful conduct of intricate and important business? Now, it has happened that such, in some cases, have eventually acquired a good knowledge of business, but it is obvious that this education has been of the most expensive kind, and what is particularly important, it has been at the expense of the proprietors, who paid a salary while the incumbent was obtaining the qualifications that would enable him to earn it.

If in the construction of a wooden building, the proprietor would trust its care only to a carpenter, who had the reputation of proficiency in his art, why should the intricate affairs of the track, machinery, workshops, and appendages of a railway be committed to men who had no previous education in the skill and arts required to be practised? Certainly there is but one excuse that can justify such a proceeding—namely, that the properly educated man could not be found.

CHAPTER XXII.

OPERATING—SUPERINTENDENT.

IN conducting the operating affairs of a railway, the superintendent has the general oversight of all departments of its business. This at least is the American plan. It has the advantage of maintaining system, and the proper harmony and efficiency of every department of the service. This is entirely consistent with the efficiency and responsibility of those having the care of separate branches of duty.

The more especial duty of the superintendent is to arrange the running of trains—establish time-tables—see that the trains are regularly run—that station duties are well administered—to establish rules and regulations for the government of all charged with the conduct of train or station duties—see that full statistics of traffic and expenses are prepared, so as to form an intelligent basis of busi ness—obtain information of the sources of traffic and its productiveness in all its particulars, so as to be able to judge of its profitableness, and whether in any respect it fails to afford proper remuneration; and under the advice of the board of managers, to establish the tariff rates for freight and passengers. It is especially his duty to study carefully the statistics of expense, in the several branches, so as to be able to judge as to the most economical method of conducting the traffic, and thereby be able to establish such rules and regulations as will produce the best result to the proprietors. In the numerous and often complicated duties involved in railway management, this will call out the best capacity and the most diligent application of the superintendent.

In some cases, the president of the railway company acts as superintendent; a method that may answer on short or unimportant railways, but cannot be recommended for other cases. The president should be the organ of the board of directors, and so far as he gives attention to the operating department, it should be as the medium of the board, and the general executive head of the institution, in which capacity he would confer with, advise and direct the superintendent. The two officers should be kept distinct, as admitting a better organization, with salutary checks, and especially as between those general officers appointed by the board of directors, and those appointed by the superintendent, which is important in securing harmony and efficient coöperation.

The superintendent should appoint or select for employment all persons in the operating department, except the heads of important subordinate departments, who, though to some extent subject to his general direction under the by-laws and regulations of the company, should be appointed by the board of directors. With the power to appoint, he should have the power to dismiss any that may be delinquent in duty. This is indispensable to the maintenance of discipline, and the exercise of proper responsibility.

On railways having important business connections, forming a general line with other railways, there will be much to demand attention from the superintendent in arrangements for the running of trains, and for tariff rates, for the common traffic of the line in connection. This duty is often embarrassed by the conflicting interests as to rates of speed as well as rates of tariff, and requires much experience and sound discretion in its management.

In all arrangements for special rates of traffic, it is the province of the superintendent to direct and control proceedings—a duty that is often difficult as it is liable to conflict with the regular rates, and to give rise to dissatisfaction among the customers of the company, and is in danger of interference with the regular checks of business. So far as practicable, it is best to avoid special rates; still, there may be items of traffic that will afford a beneficial result, and can be commanded in no other way.

It is obvious that the superintendent has charge, especially, of much that can only be occasionally under his personal supervision. The trains are, at different points, scattered over the whole length of the railway, depending on their conductors and men under their charge. The station agents are equally removed from his eye, and he must depend mostly on the fidelity of reports, to know how business is conducted. The extensive use of the telegraph, greatly aids the superintendent in obtaining information of what is going on, at distant points, but there is still much that he can rarely know. Many improper things may occur on the railway and still the train come in in time, such as fast running between stations, and particularly with freight trains, and the time thus gained improperly wasted at the stations; some one will know the fact, whose duty it may be to report it; but the repugnance of one employee to report the breach of rules by another, is such, that the report will often be neg lected, especially if it be of a nature the omission of which is not liable to exposure. Hence, the most important qualification in a superintendent is, the ability to discriminate character, and thereby place in all positions men suitable for the duty, both in regard to capacity and fidelity. In any event he must largely trust them, and his deportment toward them should be that of confidence, until he has evidence that they are not worthy, and then having carefully ascertained the fact, he should at once dismiss them, *without parley.* Some men are so distrustful of others, that their manners toward them plainly indicate that they regard it necessary to watch, and that they have no confidence beyond their personal vision. This is an error that operates unfavorably with men who know that it is indispensable to trust them. Honest men will not object to any proper scrutiny, and a discreet watchfulness may be maintained in consistency with a liberal confidence; whereas, if this necessary trust is coupled with manifest suspicion, it tends to repress the laudable ambition of honest men, and to destroy the interest they would otherwise exercise for the prosperity of the institution.

It must be confessed, that a large proportion of the men who seek and obtain employment on railways, are deficient in those qualifications that are necessary to secure an intelligent and faithful discharge of their duties; and in his selections, a superintendent is liable to be more or less deceived. But, notwithstanding this difficulty, there are many capable and faithful men to be had, and often may be selected with little doubt; and if he exercise a discreet diligence, in removing the deficient, and promoting from lower ranks of the service those more faithful, who may have developed their capacity and fitness, by filling well the duties of an inferior rank, he will eventually secure, what is of great importance to the proprietary interest, a body of reliable men to fill all places of trust. To see the proper bearing of this duty, and be able to judge well of the fitness of men for the various positions they are to occupy, is the high qualification of a superintendent; and without great tact in this branch of his duty, he will never succeed, no matter what personal capacity and vigilant industry he may exercise. I am well aware this is a difficult duty, and one in which few men succeed well; but in order to attain that success, the superintendent must have a single eye to the interests of the institution. He must be free from that vanity, that lifts him above the work he has in charge. He must know how to treat men who are worthy of his confidence, and not disgust them with supereciliousness—but be frank, candid, upright and respectful to his subordinates, that they may feel that he considers them men. He must discard favoritism, especially toward family friends, which is a baneful influence, producing dissatisfaction and the most serious evil to discipline, and consequently to the interest of the proprietors. He must, by his proceedings, encourage confidence in his firmness and discretion, to advance, as opportunity may arise, those in the service of the railway whose tried conduct and capacity may merit it. Is this too exacting? I think not. The position is evidently one of trust, in which trust must be transmitted to others, under circumstances involving important interests, and if the head of this department is deficient in these necessary qualifications, matters must go on more or less imperfectly, and to this extent the proprietors must bear the loss.

It will be seen by the preceding remarks, that the duties of the superintendent are of a highly responsible character—surrounded with cares and difficulties that will demand an upright mind—a clear head—unceasing vigilance, and business qualifications of a comprehensive order. Upon his ability and fidelity the proprietors must materially, if not mainly, depend, for an efficient and economical administration of affairs.

CHAPTER XXIII.

OPERATING—SUPPLIES OF MATERIAL.

THE supplies are a large source of expenditure in the operating business of a railway. The work cannot be long in operation before rails, chairs, spike and sleepers will be wanted for the maintenance of the track. The repair shops will require iron, steel, copper, wheels, axles, tyre, and numerous other articles, mostly hardware. Oil for light and the lubrication of machinery, cotton-waste, and fuel for the engines. On important railways it is usual to establish a store, with a storekeeper in charge, to receive and distribute supplies. Supplies are in some cases purchased by the storekeeper; in others a purchasing agent is charged with this duty. In either case the party purchasing should not pay, but certify to all bills, and transmit them to the paymaster, who, on payment, charges them to the storekeeper. In order to understand what supplies are necessary, estimates are made under the direction of the superintendent, of the kinds, quality and quantity of articles that may be wanted, once in six months, and oftener if necessary, for the guide of the purchasing agent. It is the duty of the storekeeper to receive and take care of the goods, and to deliver them as wanted to the order of such agents as may be authorized to draw supplies from the store; and to keep the accounts, so as to show the consumption by the different departments. Notwithstanding the existence of a purchasing agent, it will be necessary for the superintendent to give so much attention as will secure articles best adapted to the uses for which they are wanted, and to see that they are obtained on the most favorable terms for the company; his attention will be especially required in procuring rails, wheels, axles, tyre and oil. Cross sleepers, lumber and fuel are usually procured on the line of the railway, by agents under his direction. By whomever purchased, all articles should go on the storekeeper's books, and be there charged to the proper account, in order that his statistics may show a complete account of consumption in each department. Whoever be the purchasing or contracting agent, he should be under the direction of the superintendent, who should cultivate a knowledge of the quality and market, that will enable him to judge of the propriety of the purchases made, and guard against a quite too common vice of bribery commissions paid to the agent, to secure a high price, or a good price for a poor article. Detection in so gross and degrading a practice, should be followed by quick retribution.

It rarely happens that sufficient care is exercised in the fuel department. As yet, wood is the fuel generally used in locomotives of this country ; and much care is required in its inspection, measurement, and in seasoning. Careless supervision in the former, and neglect of the latter will be productive of much loss. Unseasoned wood for fuel, or that kept on hand too long a time, is unprofitable for use. Wood cut in the winter and piled in the open air, will generally be fit for use the August following, and should then be housed, and will be good for a year from that time, when the succeeding lot for the next year may follow; but if left exposed to the weather, it will depreciate, and become less valuable according to the time of exposure, and if it cannot be housed, no more than one year's supply should be provided at one time: being an article of general use, it is often exposed to depredation, or misappropriation. It may seem too much a detail, to require the agent who has charge of the wood at his station to account for its use; but it is the only way to maintain a check on the purchase and consumption. To do this some method must be adopted to measure and account for the quantity taken for use by each engine. This, if properly arranged for, need not be attended with much labor. Let there be racks of suitable size, placed on hand-carts, and containing a measured quantity, say one-sixth to one-fifth of a cord, not more than may be readily moved by one man on a platform. These racks may be loaded by the wood-men, and arranged in convenient position to be dumped or thrown on the tender of the engine; and on a suitable blank, fill out the number of racks, the name or number of the engine, etc., and as often as necessary return the memoranda to the agent, and he will have the elements for a daily or weekly report of the quantity used by each engine. If the platform be raised to the proper elevation, these racks, being on carts, may be dumped into the tender, and one man do the work of half a dozen, in less than half the time employed by the prevalent custom; it would require the wood to be elevated a trifle higher to facilitate dumping, but this could be done by the wood-men, while waiting for the engine, and save at least half the number required to be on hand when the engine arrives; while the fuel, on arrival of the engine, would be promptly dumped from the cart-racks, saving more than half the delay of the engine in receiving fuel, that arises from throwing up, stick by stick at a time, as usually practised. If coal be the fuel used, the same arrangement may be modified to its requirements. A system of this kind, well carried out, would not only afford a proper check on the purchase and consumption of fuel, but also a complete basis to show the fuel of each engine; an item of information well worth all the care required to procure it. It would only be carrying out for fuel the same principle of check, that has been adopted in the consumption of oil for each engine. It will be perceived, that this course would afford means to determine separately, the consumption of fuel for the passenger and freight traffic, which is important to the proper understanding of the expense of each.

CHAPTER XXIV.

OPERATING—RECEIPTS.

In addition to what has been said in relation to receipts from passengers and freight, it may be remarked: that in addition to receipts by conductors (which should be a small proportion at most) and the collections made by connecting railways, the station agents will be the principal receivers of the dues of the company. At important stations, there will be required, a passenger and a freight agent; each of these will make his report to the principal office: at small stations, one agent is charged with all the duties, administrative and fiscal. All agents charged with fiscal duties should report daily, and with reports, send in their funds to the Cashier (Treasurer or the fiscal officer who may act as such). The report goes to the Auditor, from the station agent, and the Cashier reports to the Auditor the amount of cash received, who thereupon charges the Cashier with the funds, and gives the station agent credit for the remittance. This method places on the books of the Auditor the fiscal affairs of the agents of all classes, who are authorized to make collections for the railway; consequently the Auditor is a check on all accounts, and his report to the President, or financial officer of the company, shows to the board of Directors the state of their finances. This department of business should be rigidly conducted, so that a strict accountability may be maintained, and funds not be allowed to accumulate in the hands of agents; and that the Board may, at frequent intervals, understand the condition of their finances.

In addition to dues received from the traffic of the railway, there will grow up credits from the sale of old materials, of considerable importance. A practice to some extent has prevailed of allowing certain agents to exchange these, or sell them to manufacturers, in part payment for new goods. This is not well, as it tends to confuse accounts, and often to embarrass the examination of vouchers. The proper, and as I think the obvious, course to secure a correct administration, is to make it the duty of some agent to see to the collection and sale of all old materials, who should remit to the Cashier all funds received for the same, and make report to the Auditor in the same manner as in the case of dues from traffic. The old rails may be an exception, as these materials go into the re-manufacture, and the same weight of new rails is returned, at a specific price for the re-manufacture, including the new material that must be put in, to supply waste and improve the quality; and it is only necessary for the Superintendent to collect and take the weight of old rails, and the bills for the re-manufacture, and perhaps transportation, will be simple vouchers, and charge them directly to the repairs of track, the old material being still in the track in the shape of new rails.

CHAPTER XXV.

OPERATING—DISBURSEMENTS.

The disbursements for the operating business of a railway, include payments to the officers, agents, artisans, laborers and numerous bills for supplies, taxes, damages and sundry other items, that will involve expense.

The agents in the several departments keep check or time rolls of the men under their respective charges, on which it is their duty to record daily the work done. At the end of each month these time rolls are reported over the certificate of the agent to the Superintendent, whose duty it is to examine the same, and from them cause pay rolls for the month to be made out, which should be certified by him. In the same manner all bills that may be made by authorized agents, are certified by such agents, and submitted to the Superintendent, who examines and so far as he judges them correct, approves them by attaching his signature. If bills or claims occur, out of the regular course of business, the Superintendent examines, and so far as he deems just and proper, puts them in form and approves as in other cases. This latter class is mostly for damages, taxes, and incidental expenses, and has no certificate from a subordinate agent, and depends wholly on the approval of the Superintendent. It must be apparent, that on railways of considerable extent, having a large traffic, the Superintendent can know but very little of the accuracy of detail in the monthly time rolls and bills that must be submitted for his approval—still it is necessary that they should have his sanction, as on him rests the responsibility of the expenditure, and he must depend on the fidelity of the agents, and his general knowledge of the business of the railway—his familiarity with the work in progress, with prices of labor and materials, and the current wants of the several departments, to guide him in the discharge of this duty. It will be incumbent on him to take the precaution of giving to all agents who may be authorized to incur bills, and especially the purchasing agent, such instructions as to rates, quality and kind of materials wanted, as will guard as far as practicable against improper bills, and especially against a practice that has, as before observed, too often prevailed, of agents taking a commission on purchases, and also guard against an abuse arising from influential employees recommending certain firms, or certain articles used, and from obtaining patents for others; certainly a most pernicious and degrading practice, from which it is supposed even railway directors are not always wholly exempt.

After the rolls and bills are approved by the Superintendent, they are handed over to the Auditor, who examines and corrects any clerical errors, and if there is anything inconsistent with the rules of the company, or in any respect in his judgment improper, he suspends them until satisfactorily explained, and then enters the abstracts on his books, after which he hands them over to the Paymaster, with a draft on the Cashier for funds to pay the same. When paid, the Paymaster returns the abstracts and vouchers to the Auditor, who credits the account of the Paymaster for the amount paid. In some cases the Superintendent hands the rolls and bills to a Committee of the Board of Directors, whose approval is necessary before payment. But this is inconvenient and attended with delay, especially if the officer of the board is not on the line of the railway, and in my judgment is not as well, for no man is so well situated to understand the whole subject as the Auditor, who keeps the books and readily acquires a better knowledge and understanding of the details of expenditure and the wants of this department than any committee would be likely to possess. The Auditor should be a man of capacity, firmness and fidelity, and with proper rules, established by the Board of Directors, there would be small probability of errors. This course would not supersede an examination by a committee of the Board, who should occasionally take up and investigate these and all other fiscal affairs of the institution, and so far as this relates to the operating department, the office of the Auditor will furnish the books and vouchers, and such incidental aid as a committee may require. On this point it is to be kept in view, that committees are usually composed of members of the Board, who have their own occupations, and can rarely give the time required for a thorough examination. By the system here recommended, the Auditor holds no funds; but he has the entire books and accounts, by which are established complete checks on all the fiscal officers engaged in this department, and no great wrong can occur without collusion. An intelligent board of directors will see the importance of selecting a man for Auditor to whom, from his character for capacity, firmness and fidelity the interests of the proprietors may be safely committed.

A rule of the company should establish the

principle, that in all examination of bills, full detail of items, with price and quantity, should be presented. A practice has more or less prevailed on some railways, of charging in lump several items, and especially in matters of personal and incidental expenses. This is very liable to abuse, and is wholly inconsistent with the systematic thoroughness of business that should be required for railway management, and should not be tolerated. Excuses are often made for the practice, but they should be wholly discarded.

CHAPTER XXVI.

OPERATING—STATISTICS.

In most of the States, annual reports are required from each railway within their limits, making it imperative on the officers to present nearly all the statistical information that is wanted; but this is not always complied with with that exactness of fact and detail that is designed, and, if up to the legal requirement, does not secure all the information it is desirable a railway company should have. It is often regarded an onerous and unnecessary duty, and therefore not always fully carried out, so as to meet either the intention of the Legislature or the true interest of the railway companies. It must be admitted, however, that a large amount of useful information is obtained from these annual reports, many of which possess a considerable fullness of detail.

The statistics most important for a railway company, are those that show the working expenses in each department, and the cost of every kind of traffic. A railway without this information, may carry on a branch of traffic that is unprofitable, or which does not afford the remuneration that should be derived from it. The indiscriminate mixture of this with the profitable part, both in relation to expenses and rates, prevents a clear understanding of the matter, and so long as the business is transacted on general impressions, this evil will exist to a greater or less degree, according to the multiplicity of the items of traffic. It is more especially important for so much of the traffic as may be affected by rival lines, in which case it should be known what rates will pay, at least something over expenses. A zeal for business is always commendable, but it should never go so far as to pursue it at a loss. It is also necessary, in order to judge how far the railway may enter on sources of undeveloped traffic, that have been dormant from the want of accurate information.

To collect and arrange the statistics of a railway will add to the expense of clerk-hire; a small matter, when compared to the benefits that may be derived from reliable information, to be used in economizing expenses, and giving the best possible direction to the administration of the traffic. But this is not all the benefit of thorough statistics. It gives order, efficiency, and economy in the current transactions of business, by the checks incidental to the system, and is in this respect, worth all the extra cost.

A very important feature in the statistics of a railway is, the separation of the expenses of the passenger and freight departments. This cannot, in all respects, be exact. The general officers are the same individuals in both departments; the agents at small stations attend to the business of both. The maintenance of track is also common to both branches. For these expenses some method of estimate must be adopted, and though not exact, I know of none better than the mileage of trains. The repairs of engines, cars, and coaches of each, may be kept separate, though there is often a mixture of service by engines in both departments, which on unimportant railways of small traffic may be most convenient; but it cannot be recommended that important railways should make this indiscriminate service of engines. There is no difficulty in appropriating engines according to their power to the various services; the engines for light way trains should be exclusively devoted to such trains; they are not sufficient for express trains, and to see a heavy engine hauling a way train of one or two coaches, half filled, indicates great want of skill and economy. The engines designed for express trains must be more powerful, according to the weight of trains, and will require a greater expense for repairs than those for low or medium speed—the same may be said of coaches, which must be stronger built, and of consequence of greater weight and subject to greater expense for repairing when used for high, than for low speed; thus so far as the question of separation of expenses is affected by the engines and coaches, it is practicable to ascertain it; and it is necessary to an intelligent administration of business, to know what these differences are. They can only be known by a strict attention to the statistics of expense. It is not to be supposed one set of coaches would be made for low speed and another for high speed, on the same railway, but that the new and strong coaches would be taken for the express trains, and those impaired by age and use, would be taken for the way, or low-speed trains; a course that would be dictated by ordinary business prudence.

As before observed, there are items that belong to this question, that can only be obtained by estimates—as the maintenance of track, subject to the daily use of trains of all classes. The best way to reach this item is, by a strict attention to the comparative influence of speed on the maintenance of engines and coaches; as the ratio of these will not differ materially from the influence of speed on the track. The fuel account may be easily kept separate from the passenger and freight traffic, by adopting a system of accounts to show the quantity used by each engine, as mentioned under the head of supplies.

With results carefully reduced from a thorough system of statistics, the Superintendent will have a basis, not of guess-work, but of fact, to judge of the speed he should adopt, and of the tariff necessary for a high, as compared with that sufficient to pay equally well at a low speed. He would be able to determine on all questions of rivalry with other lines, what could be done to best secure the interest of his own line.

There is yet wanting much of that thoroughness, required to obtain all the statistical knowledge needful. Profitable and unprofitable traffic is indiscriminately thrown together, and the value of the aggregate only is known, while the relative value of the parts are merely guessed at. The same may be more emphatically said of expenses. Now, it is not denied that much useful statistical information is obtained; but the fault is, in not carrying it to that detail which is necessary to obtain the information so valuable and necessary to the economical administration of business.

If a full and accurate knowledge of the expenses on track and machinery, and of the extensive, and in a great measure irresponsible, outside and inside agencies involved in rival efforts to control business—were well understood, there would doubtless be great modification in the operations of competing lines. The general practice on American railways has been to charge the same passenger fare on fast as on slow trains; a course that cannot be justified by the expenses, and must rest on some other basis: this is mostly competition, and the only one that can justify it.

There is a general opinion among railway managers, that high speed is more expensive than low, and this is the result of an impression derived from the fact, that increase of speed has been attended with increase of expenses; but the information is indefinite, uncertain, and not competent to meet the necessities of the question, or lay the foundation of business with the exactness that is attainable, and should be had for affairs of so much importance. Let it not be said the thing is impracticable; it only requires system and energy to secure the most useful results to the railway interest, and must be done before railway management can be said to be complete.

CHAPTER XXVII.

OPERATING—RUNNING OF TRAINS.

The running of trains must depend much on the weight and character of the traffic to be provided for. It is necessarily controlled by the fact, whether the railway has a single or double track. In the latter case, the arrangement of time tables is more simple than in the former, and a double track will of course accommodate a much larger traffic, and the traffic may be done at a lower rate of expense. The greater portion of the railways in this country have only a single track, and on many of these there is quite an important traffic. It is obvious that regularity is highly important, in fact indispensable, especially on a single track, as delay in one train causes delay in several others, and is particularly injurious to freight trains, which are required to give the preference to all passenger trains. It is therefore obvious that whatever tends to delay, and thereby cause irregularity, should be carefully guarded against, and the most energetic means used in case of accident, to remove the impediment to other trains. In passenger trains, high speed is more likely to cause accidents than low speed, and great care should be used to avoid unnecessary hazard from this source. High speed is also more expensive in other respects, as has been remarked in relation to track and machinery, to which may be added the increased cost of fuel. Not only are accidents more likely to occur under high speed, but the injury will be much more severe than when they occur on trains at low speed.

The question to be considered is, what rate of speed is necessary to secure the traffic? Within reasonable limits, it may be regarded true, that the interest of the proprietors will be best promoted by establishing the lowest rate of speed. There is doubtless some traffic secured by high speed that would be lost on a low speed, and the first question is, will that be sufficient to defray the extra expense? Aside from the rivalry of competing lines, this question could be easily settled. Any speed over twenty-five miles per hour for express trains, may generally be regarded as of doubtful expediency, if the traffic is not affected by competition. An express train of this speed, making few stops, and those mostly for fuel and water, would not require a high or very expensive running time; it being premised that the arrangements for taking on supplies of fuel and water admitted promptness in this respect, and no unnecessary time is consumed at stations. Such an express would have an important advantage in regularity over a higher rate of speed, and rarely miss its connections with other lines; which would arise from two causes: first, its less exposure to accidental delays, and second, that in case of delay it would be more likely to recover the loss on its moderate time. Hence it often occurs that a journey of five hundred or one thousand miles is made in less time by the moderate express than by the extra express; arising from delays occasioned by lying over at points where connection is lost with the next link in the line.

It will be contended that the public would not be satisfied with slow trains; and it is admitted some attention is due to this consideration; but the first party in interest is the Proprietor, who has furnished the means to construct the railway, and is responsible for the conduct of the traffic. Not only does he bear the original burden of the undertaking, and the current expense of high speed, but if life or limb is damaged, a much more likely occurrence at high speed, he must pay. It cannot be denied that the proprietor must be indemnified, and if it be expedient to maintain the high or extra speed, the tariff rate should be in proportion. It is the practice in England to charge a high rate for the extra express trains, and still it is contended, I think with propriety, by English writers, that even *they* do not make sufficient difference.

It is certainly a very pleasant thing to travel thirty miles per hour including stops, on a well managed railway, provided no apprehension of danger is felt; and no doubt most people would prefer it to a train of more safety at twenty-five miles per hour. But if a higher rate, corresponding to the increased expense was charged, it would be found that most travellers would be satisfied with the twenty-five mile express. In regard to speed there is one thing that may be controlled by the Superintendent, namely, that no unnecessary time be consumed by stops at stations. The time wasted at stations must be made up by extra speed between them; and a great deficiency of promptness in this duty may often be observed, and a striking difference on different railways. In one case the passengers are duly notified, and ready to leave the coach the moment the train stops; in the other, they receive no notice until it actually comes to a stand, and then it often happens that the name of the station is announced in such a confused, inarticulate sound as to leave the passengers in doubt, if not utter uncertainty, as to what station they have arrived at. The time required depends on the number of passengers to be discharged and received, and this is easily understood. Passengers are generally ready to do their part, and if it is understood there will be promptness on the part of the conductor, they will move at the earliest moment in getting off or on the train. If fuel and water are to be taken on, there will be necessity for a longer stop; but even this is often longer than necessary, and sometimes appears lengthened out to give time for refreshments that are quite unnecessary. Certain stops should be arranged to give time for refreshments; but these need not occur oftener than once in one hundred and fifty or two hundred miles, on express trains.

It is often observed that way trains, which are understood to be trains that stop at all stations, are run at as high speed between stations as express trains. This can only arise from unnecessary waste of time at the stations, or from too high a time-table speed for such trains, which in general should not be greater than eighteen miles per hour, including stops, and if the latter be promptly made, the running speed will not be great. The time lost in stops is not simply the delay at the station; but to this must be added that required in bringing down and raising the speed from and to the running rate of time. This requires good practice on the part of the engine driver and the brakeman. The steam should not be kept on so long as to require the brake to stop the rolling and cause a sliding motion to the wheels. A light train is more easily stopped than a heavy one, and as way trains are not usually as heavy, and do not require as heavy engines, nor as high speed, they may be brought to a stand more easily than express trains.

On railways having a large proportion of heavy express trains, there will be an object of some convenience in having all the coaches of the same size for both way and express trains; but the engines for way trains may be lighter than those adapted to express trains. To see, as may often be seen, a way train with one or two coaches, containing twenty or thirty passengers, and rarely exceeding sixty, drawn by an express engine with five and a half to six feet drivers, and weighing from twenty-two to twenty-five tons, is certainly no credit to the capacity of the management. On a railway having a comparatively small amount of light traffic, or such as is best done by small trains, the practice here recommended is of less importance. If, on the other hand, there is a large amount of the total passenger traffic that is light, then it is especially important that not only the engines, but the coaches should be lighter than those required for a heavy traffic.

There are railways that do a large passenger business, a very considerable portion of which is properly a light-train business, and should be done with machinery adapted to it, and with proper arrangements of trains and machinery. Much of the local traffic would be better accommodated, and as a consequence better cultivated and more economically done, than by the more expensive mode of doing a large portion of it on express trains. There are two classes of railways, on which this is very important: First, those that enter large cities, passing through a densely-settled country, having a large local traffic in passengers, that require, not heavy, but frequent trains. It is a great error to do this business by heavy, or any express train, especially if the express is run in competition with other lines. The objection that will be urged, namely, that business will be confused by this course, and you must not have so much variety in machinery, is nothing more than a confession of incapacity for proper administration, and should not be regarded, if it be intended to maintain a thorough and economical management. The second case is, a railway having a light aggregate traffic, not sufficient to support frequent trains, even if they be light. Here, in the absence of a heavy traffic, there is small hope of meeting expenses, if heavy machinery is used. It may be generally noticed that railways of this class use the same, or nearly the same character of machinery, as those doing double, or quadruple traffic; a practice ruinous to the proprietors, and unless changed, must lead in some cases to the abandonment of the railway as a means of transport. The traffic may carry them along and pay current expenses until the renewal of rails and machinery becomes necessary, when it will appear that there are no funds for such a purpose, and the railway must be abandoned, the proprietors making the most they may out of the old rails and machinery, and the district return to the old method of transport, no doubt, much to the disadvantage of all parties in interest.

In arranging express trains, especially on a railway of large traffic, much judgment is required in determining the rate of speed. It is not always necessary to run all trains alike fast. So far as they may be induced by competition, the speed of some will probably be higher than others, and they should have few stops, so as to give all time practicable to the track, and thus keep the running speed as low as will make the time required by the competition, and best secure the object sought. Other express trains should have more time, and make stops at the more important way stations, when the business may justify such extent of accommodation, leaving the minor stations to be provided for by the way trains. This will be likely to excite in the latter case some jealousy, of the influence of which the Superintendent must judge, as the railway cannot be expected to provide accommodation, beyond the fair profit of the traffic.

It is quite proper that freight trains should give the right of track to passenger trains, and in case of accidental delays of the latter, this is sometimes quite serious, especially on single track railways, having three or four passenger trains each way daily. Of course, every practicable effort should be made to avoid such delays, and at this time, on well-managed railways, they are of more rare occurrence than formerly. Freight trains require much judgment in their arrangement, from the fluctuations of the traffic, and will require modifications as the varying seasons of business occur. Those established for daily use through the year, should be limited to the number that may be required through the season of light traffic; and as this increases, extra sections of trains will be provided for a time, and as it further increases, it may be expedient to put on additional regular trains. It is very material to the economy of management, that no unnecessary trains be run, or more than is sufficient for the traffic at the time. The passenger train must be run daily, for though at times it may have a light business, it can only change by a new time-table, dispensing with a portion of the trains, which is not usually done oftener than twice in a year; but freight is more easily managed in this respect, and no more trains should be run than necessary to provide for it. This will produce some irregularity in the employment of train men, and some other work or occupation must at times be provided for them. So far as they are mechanics, this may be found in the repair shops, and others should be employed at such work as they can do, or dismissed for the time. But it will rarely happen that work cannot be found for the useful employment of laborers. Some railway managers contend this cannot be done, and that train men must be kept constantly under employment, equal to any emergency of business, which I regard unsound and inconsistent with proper economy. If due attention is paid to the wants of a railway, useful employment may generally be found for all train-men, from trains temporarily suspended. At such times, which occur after a season of heavy traffic, the machinery will require more than usual attention, and should be put in good order, so as to be in condition to meet the demands of a returning pressure in the traffic, and the mechanics suspended from the train service, may be fully employed to aid this work. When the traffic is heavy, the machinery is kept in so constant use, that only temporary repairs can well be made, leaving the more important work to be done when it is less occupied, and just when the mechanics from among the train men may be at liberty to go to the shops.

The freight trains should be ready to start on their time with the same promptness as a passenger train. There is often neglect in this respect, arising from the impression that it is less important; but the freight should be as punctual on time, both in leaving the terminus and from way stations, as a passenger train. This is quite necessary to an economical administration of the freight traffic, and to the convenience and economy of the work of repairs on the track, and the wood trains transporting fuel for the use of the engines, which are embarrassed and delayed by the untimely freight train.

The running speed of freight trains should not exceed twelve miles per hour, except for freight that can pay extra for greater speed. This principle is not sufficiently regarded, as one that enters intimately into the economy of railway transport, and will be likely to be understood only by a closer supervision of all that affects expense than has hitherto been practised on most railways. Speed in freight transportation requires a corresponding weight and power of engine, or the load must be diminished to allow the speed; in either case extra speed will involve extra expense. Whatever be the speed adopted, the load of the freight engine may be adapted to its power; in this it differs from the requirements of an express passenger engine, in which the load cannot so readily be adapted to its power; and hence the freight engine for high or low speed may be of the weight and power that will produce the best economy of transport. In arranging the time for a freight train, provision must be made for necessary detention at the stations for letting off or taking on cars. If the train is properly made up, the station service may be as regularly and promptly done as any other; the station agent should have his side track clear, in order to remove the cars to be left at his station without delay; and the cars to be put on should be ready and in position for prompt attachment. If, as will be required for way freight at small stations, a car must be loaded with freight to be delivered at several stations, and of course unloaded by the discharge of parcels for each, the parcels for each station should be loaded in the order of the stations, so as to admit of prompt discharge. If unnecessary time is taken at the stations, it will be made up, if the engine has power, by a corresponding increase of speed between them, and this will be easy if there happen to be a descending grade, and it is no unusual thing for time to be wasted by idle habits at the stations, in the expectation of making it up by increase of speed, though against the rules. This is an evil to be carefully guarded against; and attempts have been made, requiring the station agent to report the time of the arrival and departure of trains, and any circumstances of unusual delay; but unless very closely looked after, the reporting will be defective, and the Superintendent left in ignorance of proceedings in this respect. It is no uncommon thing to witness heavy freight trains moving under a time table of twelve to fourteen miles, but actually at the rate of thirty miles per hour, thundering along the rails with their heavy cars of more than two tons on a wheel, and these resting on bits of india-rubber bouncers, or short, steel springs of substantial rigidity. Nor is it a matter of rare occurrence that such things happen, and go seriously to add to the expense of repairs on track and machinery. If run at proper speed, freight engines may be heavier than passenger engines, and be no more injurious to the track; they should have low drivers, as this tends to restrict their speed, and increase their power of traction.

The Superintendent who aims to effect an economical transport of freight, will find much to demand his attention in the management of his trains. He must study, not to see merely how large a train he may haul, or how fast he can run it with one engine; but how small a sum he can make transport a ton of freight. When this last proposition comes to be well studied, machinery will be better adapted to the rail and the traffic the road accommodates.

CHAPTER XXVIII.

OPERATING—COMPETITION.

The competition of rival lines has become a matter seriously affecting many prominent railways, and has led to many convocations of railway managers, for the purpose of harmonizing the conflicting interests. These assemblies have produced agreements as to running trains and rates of tariff, having at least the ostensible object of securing fair rates and a proper division of traffic. There are, however, some inherent difficulties in the way of such arrangements, arising out of the peculiar circumstances that control traffic at the same rates on different and rival lines, which when left to its natural action, will flow more to one than to the other, and the result is pretty sure to cause suspicion and jealousy on the part of those having the least ability to control this tendency; and when trade flows against them, the suspicion is easily aroused to account for this result, as arising from some transgression on the part of their rival, against the programme of the arrangement previously entered into. And, perhaps, without waiting to obtain correct knowledge of facts, they proceed to measures equally inconsistent with the contract, and the parties are soon floating in the broad field of rivalry. In some cases of rivalry, there is no material difficulty of this sort, and with fair arrangements each party will obtain its due share of the traffic.

No railway can consent that its rival should be allowed to establish lower rates for a competing traffic than its own, for of all the advantages that control it, there is none so effective as lower rates, and any rival line, possessing inferior inducements, should be content with uniform rates, and proper time arrangements for the running of trains, and take such portion of the traffic as will naturally flow to it under the circumstances that may exist. Without this, no arrangement can be made of any practical value—and it is better for the inferior line to submit to this, and obtain fair rates for what it carries, than struggle by various means to do a business that will not pay expenses. But no difficulty need to exist, where the lines will naturally make a fair division of the traffic competed for at equal rates.

It must be conceded, that in general this kind of arrangement or contract between rival lines, has not been very successful in promoting the interest of the proprietors. It has rarely happened that they have been in force any great length of time, before some party imagines he can make something out of his adroitness, and sets at work to obtain some advantage not consistent with the contract; he may employ agents to solicit traffic, and secure an influence by misrepresentations, which are not particularly unusual with that class of men, who often exercise great skill, and produce considerable effect on the course of traffic; or, by the aid of free tickets, drawbacks and commissions, the rates are often actually reduced, though the tariff stands formally as unchanged. Such practices are sure to be found out by the injured party, before a large amount of traffic can be secured, and of course measures are taken to counteract them, and the agreement falls to the ground. The measures that have been attempted to enforce such agreements, and punish the faithless party by fines, have not produced any important results that I have known. This class of contracts must rely mainly on the good sense and fidelity of the parties.

Railway managers make great efforts to establish arrangements, that will enable all to obtain fair rates and reduce expenses, a proceeding that should be beneficial to their mutual interests; but the several parties (usually the principal operating officers), pursue this with different motives; one enters upon the negotiations in good faith, intending to carry out faithfully the arrangement he agrees to, while another only designs to make it a cover, to secure undue advantages. No sooner is the contract made, than the latter, considering his rival restricted by the contract, and relying as before observed on his ingenuity and cunning, goes to work to defeat the object of the contract by various devices, which he designs to hide from his rival; and by employing the means before described, aims to influence the direction of traffic, and all in conflict with the contract between the parties.

In view of the short time that such measures may be practised, before the facts will come to the knowledge of his rival, the practice indicates great want of sagacity on the part of the offending party, whose bold denials and slender evasions, to fasten the fault on some subordinate, are often as ludicrous as they are dishonest.

It is quite apparent that contracts with such parties must be of short duration, and end in open rivalry, greatly to the prejudice of the proprietary interests of both lines. Some operating managers appear to regard this as shrewd practice, and as a laudable aim to advance the interest of their lines;

but it presents a grave question to their employers; for if their agent is ready to cheat their rivals, little doubt can be entertained, on the occurrence of an opportunity, that he would be quite as ready to cheat them. The real difficulty is, the employers generally know very little about the matter, and what they do know, is usually from the offending officials.

The mischief of competition cannot be healed without good faith on both sides. In the absence of this on either side, all efforts will be abortive. But the end to be secured, namely, fair rates and moderate expenses must be reached, or the proprietary interest must suffer, and the means suggested are indispensable from the circumstances of competing lines of railway. Certainly so far very little has been effected in securing fair rates for rival traffic, a fact by no means creditable to railway management.

It is in the nature of things, that railways holding certain relations to each other, divide the traffic, or a certain portion of it, on equal tariff rates, and the traffic must take one or the other line. The question for the proprietors therefore is, shall equitable arrangement be made and maintained, which shall secure fair rates and economize expenses? or shall we, by costly, and irresponsible agencies and other extra expenses, endeavor to show large gross receipts, and thus control at unremunerating rates a traffic that should afford a reasonable profit? It must be borne in mind that no increase results from extra expenses and low rates, for the rival line, by similar efforts, still divides the traffic in about the same ratio as would be the case on a fair and equitable contract. The public gain by the low rates, and under this influence there will probably be an aggregate increase in the volume of the traffic, but this will be lost in density and value, and the proprietors will only have the consolation of loading their track with a traffic that yields little if any profit, if not productive of positive loss.

In the condition of many railways in this country, the amount of traffic in competition is very large, and the proprietary interest involved in its management very important. Cases occur in which it is very difficult to manage it successfully; but in many others, the only thing necessary, is a sound business sagacity, exercised with fidelity and firmness to secure paying rates. A competing railway is not like the competition of stage coach lines, where one party aims at breaking down the other, and thereby securing a monopoly of the trade; for a railway that has a local traffic sufficient to keep it in operation, will always have the power to act on its rival. It cannot be broken down, and will always be in the field as a competitor, and if it cannot maintain the rival traffic at a profit, will be able to prevent its competitor from making profit; and though naturally inferior in capacity, its inferiority is not often so decided as to prevent it from destroying in a great measure, if not wholly, the profits of the rival line.

It is sometimes urged that a railway is so loaded with capital and other liabilites that it cannot compete with one less burdened in this respect. A little reflection will indicate the fallacy of such reasoning; for whatever may be the liabilities, if there is sufficient local traffic to pay current expenses, and so much for the use of the rails as they are worth for old iron, it is pretty certain to be kept running, and there are few that cannot do this on their local traffic. If the railway was a project to be made, then its cost would be a legitimate question in reference to its ability to maintain successfully a competition with a line already in operation; but it is constructed, and whether it has cost little or much, will most likely find traffic enough to perpetuate its running, and keep it in the field as a competitor with lines, having in some respects greater advantages. Is it therefore wise to lightly esteem a rival?—far better to arrange terms on the basis of existing facts, and by equitable arrangement secure a fair rate for the traffic to be carried.

CHAPTER XXIX.

OPERATING—FINANCIAL MANAGEMENT.

IT would be superfluous to discuss in detail the leading object the proprietors have in advancing funds for a railway. A limited number have doubtless been influenced by considerations of the indirect benefits they may derive from the results of the anticipated working of the railway on property and trade; but in the main, they look for remuneration in direct net profits from the traffic. It cannot be doubted that the indirect benefits are far greater than the direct, and these accrue, not only to those who subscribe with this object, but to the community on the route, generally, who may not, and generally do not, furnish any material portion of the funds required for the work. For the main part, the funds have been furnished by those who take the securities as an investment, depending wholly on obtaining compensation from the net profits of the traffic, and are no way interested in the indirect benefits that result to property and trade. This class, and especially those that rely on the stock securities, take the financial burden of the railway. The class of stock subscribers relying on the indirect advantages are usually quite small in proportion, and are pretty sure to find their full remuneration.

The stock proprietors, having usually waited some time for the construction of the work, naturally look with anxiety, as soon as the railway is opened for traffic, for the prospects of a dividend on their stock, and it is quite proper the Directors should feel a desire, so far as they may have power, to gratify this reasonable anxiety of the stock proprietors. Few railways have been completed without contracting a debt, with a prior lien on the property, that must be provided for before dividends can be declared, and the first appropriations from the net income on traffic, must satisfy the interest and sinking fund on the funded debt. In the incipient operations of the railway, this funded debt, according to its relative magnitude, will interfere with dividends to the stock proprietors until the net income on the traffic shall equal the same interest on the whole capital that is paid on the bonded debt, with a surplus over this equal to the sinking fund, and then both classes of proprietors will be equally well paid for the time. For the time that may elapse before this result is reached, the bond proprietors will be fully paid, while the stock proprietors may receive a small, or perhaps no interest on their outlay; and can only hope an increasing traffic will eventually afford them an equivalent for the loss of back interest. If the funds were all from stock proprietors, and no bonded debt incurred, the whole net income would be applicable for dividends to stock proprietors, and dividends would be earlier expected. This, however, is not often the case; a bonded debt, at least equal to the amount of stock, is more common, and to provide for the interest and sinking fund on the former, is on many railways an important, and sometimes in the early operations, a difficult duty in the financial management; but as the funds obtained for the bonded debt, could only have been obtained on this basis, there is no alternative, unless, by some evasive repudiation, it is neglected for the benefit of the stock proprietors; a process of cheating that can only be classed with other acts of infidelity, whereby men appropiate the property of others to their own use.

At the opening of a railway, the net income must be matter of conjecture, and considerable time will be required to ascertain it with reasonable certainty, and thus determine the basis for dividends on the stock. It usually happens that conjecture is sanguine, and urgently presses the promise of early and liberal dividends; and it becomes important to look carefully at the bearing of this question on the permanent interests of the proprietors.

On the opening of a railway, there is, in most cases, a respectable amount of traffic ready for its occupation. This will increase by the well established laws of trade—namely, that improved facilities enhance the development of existing traffic, and new items of trade are brought out by the superior cheapness of transport. If the natural resources of the district accommodated by the railway are but indifferently occupied, as may be peculiarly the case in a newly settled country, it may be expected that the increase will be large, as compared to that enjoyed at the opening, and though dividends may be small, or even nothing in the commencement of the traffic, they will eventually be remunerating.

The railway and machinery, if well constructed, being new, the repairs and maintenance for a few years will be comparatively small, and the net profits will, for the time, be larger than will ultimately be found an average rate on the traffic. Continued use will impair the track and machinery, and new parts will be required for their efficient maintenance and usefulness, consequently it must be expected that there will be an increase of current expenses in the operation of a railway. It is usual to estimate the probable increase in the traffic to be not only sufficient to meet this increase of expense, but to even more than do it, and to justify the expectation of an increasing rate of dividend. The more prudent course would be, to make adequate provision by a reserved fund, that may very properly be appropriated to the purchase of new machinery, to meet the growing wants of increasing traffic; instead of the too common course, of increasing the indebtedness of the Institution for this purpose. The latter course is most likely to be pursued, if there be an influence in the Board of Directors that seeks to elevate the market value of the stock, as it is well understood that dividends are a material element in promoting such a result, without much reference to the sources from whence they are derived. If, on the other hand, the railway property is regarded as an investment of funds, and governed on the same business principle as would control an intelligent individual, who would not esteem the payment from one hand to the other as evidence that his property was thereby enhanced in value, the management will be different. All efforts to produce fictitious value, either a rise or fall, can be no benefit to the property; and can only be viewed as means of deception, practised on those who may unwittingly repose confidence in unfaithful managers. It is sometimes said in opposition to this view, that the stock proprietors are entitled to all that has been earned, and the managers have no right to appropriate net earnings to purchase new machinery, or for other purposes legitimately belonging to the construction account, and that funds for these should be obtained on loan, or by the issue of new stock shares, even if by these means funds can only be obtained by a considerable discount. It is urged, in support of this method, that the stock proprietors cannot afford to do without dividends, to the full measure of net income, and that it is not right to withhold them from them. In carrying out this principle, great skill and liberality is often practised in charging over to construction, for the purpose of showing an apparently small amount of current expenses. And hence the construction account has come to be regarded an abyss, never to be satisfied, constantly swelling the liabilities of the company. Under this mode of proceeding, dividends may be made, provided the managers have credit, or the Institution has credit to borrow funds, until the means of dividend be totally exhausted. The test of this argument, or method of business, will rarely, if ever, be made by the stock proprietors, so long as they receive the semi-annual dividends of four or five, or even three and a-half per cent.; and they will not often inquire as to the source from which the dividend money came.

Men who have invested their funds in railway property have the right to all the property can earn, and no doubt many have need of early income, and whether they need it or not, are entitled to receive it, so far as net income has been produced by the traffic; but their interest as an Institution certainly will not be promoted by borrowing funds, instead of using their own, to meet expenses that must be incurred to extend or sustain their works. It may not be best to dispense with early dividends, and if affairs have been discreetly managed this will not ordinarily be necessary; but the dividends should be made with a view to provide for depreciation, and the funds reserved for renewals need not be idle in waiting for depreciation to occur. They can be used, as before stated, in providing new machinery for increasing traffic, instead of funds being borrowed, for such purpose, at a discount, and these may be charged to construction, and remain in that shape as a surplus account.

In order to illustrate the method recommended, let it be supposed, that the strict net income, after the payment of expenses, interest and sinking fund, shows a balance of eight per cent. per annum; which may be the case on a comparatively small traffic, on a new and well made railway while expenses are low. Let five per cent. per annum be paid as dividend to stock proprietors, and three per cent. be carried to surplus account. This surplus will probably be needed, for additional machinery, for side tracks, ballasting, or other necessary improvements that may be demanded by an increasing traffic. By this plan the surplus will be safely invested in property required by the wants of the institution. The second year the improving traffic we will suppose adds to the net income one per cent., and if not called for to meet any greater demand than was made the previous year for extensions and improvements in the works, let the dividend be six per cent. per annum for this year, leaving the surplus of three per cent. to be carried forward in the same way as before. After this, as the traffic may warrant, pay six per cent. dividends and no more, until the surplus can no longer be discreetly expended for necessary objects of construction. If the income continues to improve, as it most probably will, and the net earnings warrant it, let the dividend be raised to seven per cent. per annum, and so far as the surplus is not wanted to increase the machinery or improve the works, let it be appropriated to purchase the bonds of the company. When the bonds and all other indebtedness of the institution is liquidated, it will be proper to convert the surplus account, as circumstances may warrant, into a stock dividend, to be issued to the stock proprietors, who will now be the sole proprietors. The railway being now in good condition, with ample machinery for its traffic, and the interest and sinking fund account closed, it will be able to make its regular dividend on a corresponding increase of capital, and the stock proprietor finds his compensation, and though deferred, his investment proves the best in the institution, as it ought, he having taken all the risk of the enterprise. If he has not received the benefit at an earlier day, he finds no serious losses by discounts to obtain funds, and, by a judicious management of the surplus, he finds a property in sound condition as to its works and outfit for its now enlarged traffic. There has been no demoralization of its operations, from insufficient means of prompt payment, and he now gets back the full amount, with interest, of all that was due from net resource. On this method the rate of dividend in the early stage and the advancing progress may vary according to the exigencies of the case, and the profits of the traffic; but it should not be high, and should depend on the wants of the works, and the amount of indebtedness, keeping in view the policy, not only of avoiding new liabilities, but of making as early liquidation of existing indebtedness as may be consistent with giving a small but regular return to the stock proprietors in the form of dividends. Such a system of financial management would be more likely than any other to induce economy and fidelity in conducting the business of the institution.

Regularity and reasonble certainty in dividends is generally regarded important in this kind of property, and especially to those who depend on it for means to meet current expenses; but trade will fluctuate, productions will be irregular, and these will cause irregularities in the traffic, and consequently in the net income of the railway. Changes must take place in this branch of business, the same as in all others, and income will necessarily vary; this must be anticipated in all systems of finance. In order to secure regularity in dividends, it is necessary to ascertain the average of several years, which will embrace the extremes of fluctuation, and by reserving something as a surplus from the more productive years, be prepared, from this reserve fund, to meet the regular dividend in those years that prove less productive. And if it appear that a larger reserve has been made than the exigencies of the finances require, and especially if there be no debts to pay or provide for, let it be converted into a surplus dividend, and carried where it belongs, to the credit of the stock proprietor. This principle of finance is the more important, if the institution is under obligation to provide interest and sinking fund on a bonded debt, or has important renewals or improvements to make in its works. If the railway be mainly, or wholly, a stock property, and the proprietors prefer to do so, it is perfectly proper for them to divide each year, according to the net income; but they must expect, on this method, to experience more or less of fluctuation in semi-annual dividends, and thereby increase the temptation to stock speculation on the part of their managers and officers, who will best know the effect of the dividend on the market value of the property.

I am well aware that the plan of finance here proposed will be regarded with disfavor by many railway managers; who prefer more of what they term liberality in dividends, and do not fear to augment and swell the construction account; a mode of proceeding that has a glittering present, and, as has often proved, a most disastrous future. The great question for the proprietary interest to consider is, What shall be the basis of financial management? Experience has shown that it must be either a speculative or a business basis. The former is a dangerous policy for those of the proprietors who may not be of the party in control to reap the benefit. All cannot speculate with profit —there must be a portion of the proprietors to be speculated upon, in order to have a basis for successful speculation, and this class must be the unsuspecting portion of the stock proprietors. To the initiated, there will appear no great difficulty, where there are numerous proprietors in a joint stock property of this kind, for cunning and unprincipled men to waste the property, at the same time they are making fair, and perhaps good semi-annual dividends. The first serious check to the course of the speculating party, will appear in their inability to make further loans; for it usually happens, that the money-lender is more acute in apprehending the tendency of affairs than the stock proprietor, who may be lulled by dividends; and the cessation of the power to loan is usually followed by a cessation of dividends. This condition of the finances of the institution will most probably be anticipated by the shrewd speculative manager, who takes care to dispose of his interest to unsuspecting parties, in time to avoid loss, while the great mass of proprietors are suddenly roused up, just in time to see their property take wings and fly away. There is another feature of the speculative policy very mischievous to the proprietors, namely, the practice of buying and selling the stock of the company on speculation, by which the parties engaged are at one time interested in a rise, and at another in the fall of the market value. In regard to this practice, it may be said, any manager has a right to buy and sell the stock of his company, and that he injures no one by so doing. This is a plausible reason, and is sufficient for all the ordinary transactions of this kind, based on the circumstances and wants of the manager for actual investment. But when the chief officers and influential managers of a company are known to operate largely in a speculative way in the stock, they are strongly tempted to neglect the proprietary interests committed to their trust, at any time when their operations in stock will thereby be promoted. And further: The men whose minds are of a speculative turn are liable, and indeed it may be said generally do, devote so large a share of their care and attention to objects and means of speculation, that there is not sufficient time and thought left to manage successfully the affairs of the proprietors, even if they made the best of this hastily snatched remainder. No doubt it may be set down as a standing caution, that if a proprietor chance to learn that the President, Treasurer, Secretary, or Superintendent, or the more active and influential directors, are engaged in large stock operations, making time sales and corners, it would be wise for him to watch a favorable opportunity to dispose of his interest in the institution. This method of business is only to be adopted by such managers and officers as prefer the chance of speculative results, and whose prospects will depend on the number and interest of those who may incautiously be induced to engage, on the supposition that affairs are, and are to be, conducted for the general interest of the proprietors.

The other—namely, the business basis before mentioned, is predicated on a system of management that seeks its results in a wise, just and business-like conduct of affairs. Practising sound economy in all objects of expenditure—discarding all traffic that will not pay some net income—having no taste for a large gross revenue that is accompanied by equally large gross expenses—nursing all traffic that can be found to pay a reasonable, or at least some, net income—in short, practising the course usually followed by discreet business men in the conduct of their private affairs; regarding the prosperity of the institution as the controlling object sought, and pursuing this with singleness of purpose, aiming to exercise the same unremitting attention and sagacity that actuates a discreet individual in conducting his affairs. So obviously just and proper is this method, both in regard to the construction and the operating of a railway, that the manager, however his individual tendencies may be toward speculations and jobs, will profess to be wholly guided by its rules; and it may not be easy for the confiding proprietor, or even his co-manager, to always understand, until too late, the unsatisfactory basis of such professions. This latter, however, is an evil incidental to humanity, and no man may be able to penetrate the veil that hides the purpose of the manager; and perhaps the purpose is scarcely formed, or if so, very imperfectly matured; and yet it will probably be developed in the ultimate prostration, if not the total ruin, of the institution. That this evil is one of imminent danger, is sadly proved by the history of railways; and the question naturally arises, How are the proprietors to guard against it? A question more easily asked than answered.

The stock proprietors, though first to feel the effect of incompetent or unfaithful management, are not the only parties to experience the disastrous results. The bond proprietor may also suffer, especially if the bonded debt be a large proportion of the capital of the institution. This class, having no voice in the management, are usually very quiet so long as their interest is punctually paid. It is very natural that they should feel a higher degree of safety than the stock proprietors, who may be wholly sacrificed before the bond-holders feel a loss. Experience, however, has shown that they are not free from the ills and hazards of bad management. It is true the stock must be first sacrificed, and after dividends can no longer be paid to the stock proprietors, the interest may be continued for a time on the bonds; but it rarely fails, that the policy which has destroyed the stock, will also destroy, more or less, the security of the bonds. Secured as the bonds usually are by a mortgage of the whole property, it would seem to be in the power of the bondholders, on the failure of the company, to take possession by foreclosure, and enter into the management as sole proprietors; but there are difficulties in the way. In the first place, there is the difficulty of obtaining concert among the bond proprietors to act in the premises; and as more or less of pecuniary responsibility must be incurred, a part do not like to put themselves under engagements, without the coöperation of the whole, and here they find themselves somewhat in the position of stock proprietors, a numerous body, incapable of that coöperation necessary to secure efficiency, and though some are ready and anxious to proceed, others will hold back, to avoid responsibility, while they are quite ready to avail themselves of the benefits that may result from the efforts of their co-bondholders. The second difficulty arises from the unwillingness of the stock proprietors and unsecured creditors to allow the railway to go into the hands of the bondholders under a foreclosure that will cut off their claims. Though no dividend is paid on the stock, it may be a well founded opinion that good management, and improved prosperity of trade, may show the institution to have some value for the stock proprietor, and this it is natural he should desire to secure. The third difficulty grows out of the floating debt, which is usually large, under such circumstances, and much of it due to laborers for operations on the railway, and for supplies in the operating department, mostly due to persons in the vicinity of the railway. This class of creditors usually threaten to mob and drive off any party that shall attempt to operate the railway without acknowledging their claim to payment. The fourth difficulty is found in the courts, that in some instances either repudiate the power of the bondholders to take absolute possession under the mortgage, or so qualify this power as to make it little more than a nullity; but if the courts find the terms of the mortgage so explicit that they cannot evade them, they can often aid in protracting the proceedings so as to work much embarrassment to the bondholders. It is not to be inferred that all courts proceed in this way, for there are yet found judges who feel the responsibility and dignity of their position, and may be relied on to maintain the object of their institution, namely, the administration of justice between parties, without regard to the parties. It is indeed unfortunate that the latter class are not more numerous, but they no doubt hold their relative position, according to the degree of our civilization, and it must be borne in mind that adjudication between a railway corporation and an individual is a thing that tests the firmness of a court in a high degree. Notwithstanding all the difficulties, as the right of the case is on the side of the bondholders, they ultimately succeed, if the traffic of the railway is worth it, in securing the payment of principal and interest, though they are sometimes induced to acquire control under stipulation to pay a certain class of operating debts, and after reimbursing their own interest, to give the stock proprietors and unsecured creditors whatever may remain, and which is an equitable result not often complained of by the bondholders. Mortgage bonds of a railway corporation, which, from inadequate traffic, or ill-management, does not pay its stock proprietors, are far from being that quiet and reliable source of revenue that many have supposed. If the management of a railway is under speculative control, that management will be exerted to sustain the stock or the bond proprietor, as may happen to be the interest of the controlling party. If the stock controls, and the prospect is regarded doubtful as to the ultimate strength of the stock, it may lead to movements that will change this interest over to the side of the bondholder, and in the midst of a cloudy and vigorous litigation, the path of right may suddenly appear so well fortified that further resistance will be vain, and matters are brought to an easy solution; the bondholders must, at least, be provided for, and the stockholders may take what remains. The bond proprietor is not at fault, he had no control in the management that compelled this result, and the stock proprietor must bear in mind, that if his property is wasted, it has been wasted by his representative, and though his trust may have been abused and forfeited, it is no fault of the bond proprietor. Instances have occurred in which the bondholders have compromised their claims, by delivering up the coupons, for one or more years, either gratuitously or taking therefor new obligations of the company, secured by a younger mortgage. There may be circumstances that render this policy expedient, as from the recent operations of the railway the undeveloped traffic may not afford the means of payment, while the future prospect is favorable for the ultimate ability to pay current interest and sinking fund; and if there is confidence in the fidelity of the management, it may be the wisest course for the bondholders. But if the condition of the company has been reduced by incompetent or unfaithful management, no good is likely to result from this method, as affairs will not be likely to improve, and when the time comes for payment to be resumed, the finances will be no better, the track and machinery in worse condition, and the bond proprietors no more favorably situated for securing their rights than they were when the coupons were surrendered. As a general rule, it is best for them to proceed to take possession as soon as payment fails; delay rarely works to their advantage, and promptness and efficiency in action are quite as necessary to secure their rights as those of others. It should be borne in mind, that the same incompetent or unfaithful management that has destroyed the capacity to pay dividends to the stock proprietors, will be likely to prey on the bond proprietors as soon as the stock interest is destroyed or paralyzed, and a new administration of affairs is necessary to secure the property. It is well and equitable for the bondholders to take possession, under stipulation, after providing for their interest and sinking fund, to pay any surplus over to unsecured creditors and stock proprietors, and very likely the latter interest will be best promoted by such change of administration, in removing from power a set of unprincipled speculators, or incompetent managers, who will never advance the interest of either class of proprietors, or their creditors. A railway that cannot pay interest on its bonds must indeed be a very poor project, and the fact that interest is withheld, may be generally regarded as evidence that affairs are not properly conducted, and will not improve by indulgence. It is admitted, as before observed, that there may be exceptions to this rule; but it calls for careful scrutiny, if the bond proprietors would secure their rights.

CHAPTER XXX.

GENERAL REMARKS

The management of a railway must necessarily be a trust committed to men, and the hope of the proprietors must rest on their sagacity in selecting such as will prove competent and faithful. In cases where this has been successful, institutions of small original promise have proved remunerating, and in others, where the speculating class have by any means obtained control, the most promising railway enterprises have been so reduced as to greatly deteriorate, if not to utterly ruin, the interest of their stock proprietors.

The first organization of a Board of Directors for the management of a railway is highly important to its ultimate success. Experience has proved that if unfaithful men obtain control in the original organization, it is very difficult to remove them, especially in large and important institutions, having a great number of proprietors, and those scattered widely apart—unacquainted with each other and the details of their property. The necessary combination required for that efficient action which is requisite to effect a change, is hardly practicable, at least until the institution has fallen so low in business and credit, that dividends can no longer be made—the interest on the bonded debt deferred, and an over-shadowing floating debt leaving little for the stock proprietor to hope for. Nor is it necessary to this result that all the Board of Directors are unfaithful to their trust. A clique of three or four out of a dozen directors may so manage that their associates will be kept mainly in the dark, and if any hitherto confiding member should chance to get his eye on some suspicious matter, and venture to press inquiry too close for the ruling party, he is soon branded as an old fogy, quite behind the times, and his unsuspecting colleagues will learn that he is impracticable and cannot comprehend the age, or their business, and if he continue to be troublesome, he must give place at a future election for a successor more easily controlled. On the other hand, if there be three or four good men in a Board of Directors, who take active interest in the affairs at the outset, with a President as the executive head who has the necessary capacity and fidelity, they will most likely conduct the management beneficially for the proprietary interest. There are usually several members in a board of this kind who institute very little scrutiny into business, and generally act without much question on the measures proposed by the ruling party. If these latter are fair men, they will respond to the arguments that aim to promote the interest of the institution; and under the lead of the President and three or four correct men, who carefully explore affairs, there will usually be little danger to apprehend from a speculating clique, and this organization may practically be equivalent to a good Board of Directors, even if there be several speculating men among them. But it is best to keep clear of the latter class, as their personal schemes will ever require watching, to prevent the indirect waste of the resources of the company. It is moreover hazardous, as competent and upright men who may be elected, perhaps with the sole view of giving character to the institution, do not always expect to give special attention, and may not be aware that the board has members that require watching, and therefore the safe policy is to secure a full board of good men, who will uprightly pursue a policy designed to promote the interest of the institution.

It is not generally expected that all the members of a Board of Directors will examine with particular care the detail of affairs; this is left to certain men who are relied upon to do such duty, and more particularly to the President, who should be the executive head of the company, and on whose fidelity and capacity the prosperity of the institution will in a great measure depend. At the same time it is very important for him to be supported by intelligent and upright associates in the board, or there will be danger that he may be baffled by adverse interests, and though holding a position of ostensible responsibility, his plans may be defeated by various misrepresentations and stratagems, that for a time he may not suspect, nor until their development will it appear that his position is only nominal, and that he does not possess the power to make himself useful to the proprietors.

If an institution of this kind is so unfortunate as to have in their Board of Managers a clique who aim to make it subservient to their individual rather than the proprietary interest, it is likely that they will contrive to get the finances under their control, and place the various departments of business in the hands of their relatives, or other personal friends, who will understand just how far they may go in carrying out the purpose of the President, or party opposed to their proceedings. Measures adverse to the proprietary interest, if there be one or more persons in the Board who aim to be faithful to their trust, will not probably be at first manifest. They will proceed gradually, and perhaps quiet some unsuspecting member, that otherwise might not look on with indifference, by a bit of what he may suppose a harmless *realization* in some projected operation, ostensibly of fair business aspect; and it is surprising how many men of fair standing among their neighbors allow in this way an embargo on their actions until too late to remedy the evil, and so the mischief is fastened hopelessly on the interests of the proprietors.

It has been suggested that men who aim to manage affairs of trust for their individual, rather than the proprietary benefit, are not likely, and indeed are not generally good business managers, and fail to make the business of the institution as profitable as it should be. On this point it may be set down as a rule, that it is not usually their interest to have in their service a very scrupulous body of agents; but rather their personal friends, who will, from affinity and interest, be well disposed to carry out their wishes, as more important than a competent and faithful discharge of their respective duties. In this respect, family connections are likely to be sought, as more confidential and reliable for their purpose, and also from a disposition to pension on the institution members of the family, who may not be able to otherwise provide for their support. This sort of management is destructive of discipline, of honorable emulation among other employees, and all generous zeal for the interest of the Institution. It is not necessary to be exclusive on this point, for there are, no doubt, cases where a relative may be advantageously placed in position of more or less responsibility; but it should only be done where there are manifest qualifications, namely: a good knowledge of the duties to be performed, practical sound sense, that will fully meet every question of discipline, and vigilant application and fidelity in the discharge of every duty. The reverse is the more general result of this application of patronage. Men occupying positions through the influence of friends in power, are likely to depend on such influence rather than on a proper discharge of duty, and are often prone to assume a consequence subversive of the discipline, harmony, and efficiency of business. The evils of nepotism have usually been mischievous to railway institutions, and any indulgence in this sort of patronage should be looked to with the most rigid caution, as it is most probable that managers and general officers who adopt it are not particularly scrupulous for the interest of the proprietors.

To further schemes of speculation, it is often regarded important to show statements of large gross receipts. To effect this, rival lines reduce rates below a compensating tariff, incur large expenses for agencies, many of them of loose and irresponsible character, and wear out the rails and machinery to keep up a delusion that can only result in eventual damage to the proprietors, though for a time screened from their view by a free use of that absorbing and *ruinous abyss*, *the construction account.*

Branch lines are expensive means of forming connections with other railways, and have often proved very useful to some parties in the management, while they have been more or less injurious to the interest of the proprietors. There are, doubtless, cases where proceedings of this kind have proved beneficial to the interest of the main line; but experience in this and other countries has shown that local interests, contractors, and jobbing managers are the parties most usually benefited, while probably, in nine cases out of ten, the proprietors of the main line have suffered loss. It should be borne in mind that there are very few branches that can pay, either directly or indirectly, for their cost. It is not to be supposed that operations of this kind are in all cases chargeable to dishonest management; they are no doubt entered upon, in some instances, in the honest expectation of promoting the prosperity of the institution; but they are very liable to be the occasion of a different purpose, and therefore, any proposition of the kind should be thoroughly investigated by the directors before they are committed to the expenditure. In this matter they will meet with many plausible representations, both in regard to branches and aid to extending lines, from parties interested in their advancement. These parties will generally have, or claim to have, ample local information as to facts and estimates, and consequently an advantage over others in the argument, and the only safety is in taking full time to consider and scrutinize all questions involved in the proposition. Delay is generally the prudent course, and the disaster that has so generally followed this policy should certainly be sufficient to protect against hasty action, or until after the most careful consideration of all the circumstances of the case, that may bear on the interest of the institution.

It is sometimes very boldly and plausibly urged, that directors should have a large interest in the stock of the company, as an inducement to give close attention to their duties; and no doubt this is an important consideration. Directors usually receive little or no compensation for their services, and if they have but a small proprietary interest, there is not an object of sufficient importance to induce much attention. Some directors, having very little private occupation, accept a seat in a Board, as a means of social intercourse, or of honorable trust, and do not expect much labor. Others, with more or less of proprietary interest, are men very much absorbed in their private business, and are restless every moment they are occupied in their duties as directors, and do scarcely more than to vote, or approve of whatever is laid before them. With a large pecuniary interest in the institution, it is reasonable to expect a more close attention; but this of itself is not sufficient to protect other stockholders, and it is indispensable to their safety, that the directors be not only capable, but upright men, who, in looking after their own interest, will do so with strict reference to that of all other proprietors; and it should not be forgotten, that a director with a large stock interest, may also have other large interests to be affected by the operations of the company, or plans of speculation that intensify his other interests in such a degree, that his proprietary interest in the institution may be quite secondary.

In many cases it is no doubt a difficult thing for the proprietors, who find their affairs in the hands of a board of doubtful integrity, to elect a good and reliable board in their stead. Whether this is more difficult now than it was formerly, it is not necessary in this place to discuss. It may be stated as a general principle, that the successful management of all joint stock property requires a high degree of civilization; and this is peculiarly the case in railway institutions so situated as to require large capital and numerous proprietors, mostly strangers to each other, and to the detail of their affairs. It is apparent that the management of their affairs is eminently a matter of trust, affording very little opportunity for an intelligent scrutiny. Adequate capacity for administration and fidelity to so important a duty, can only be expected to result from a high civilization, in which a conscientious sense of *right will constitute enjoyment far outweighing that of any gains secured at the expense of dishonest proceeding.*

It is a question of grave import, whether our civilization has reached that degree that can give reasonable confidence in this kind of property, especially to small proprietors. No doubt it will be claimed that it has. It can hardly be expected that the now numerous class who occupy these places of trust, will acknowledge any deficiency, and though there are doubtless many upright men in such positions, there is still no room to doubt that the railway interest, both in this and other countries, has largely suffered from the low standard of civilization prevailing with both them and us.

The management of an important railway involves arduous and responsible duties, calling for large business experience and qualifications in judging of men and things, and the most competent and upright directors will find ample scope for the exercise of their best faculties. With the most upright intentions they will probably commit some errors. The first and most important duty is, to select from their number a man to preside over their deliberations, and to take the administrative charge of the executive department of the affairs of the institution. It is the most usual course in the election of a Board of Directors to have in view one of the number for this position. Upon this officer, as before observed, the prosperity of the institution will greatly depend: not, however, as is often supposed, by his vigilance and attention to mere details—though on a short and unimportant railway he may consistently give much attention to details, and perhaps save the duty and expense of some other agent—but on what may be termed important railways, his attention should be directed to the general interests of the institution. and especially to see that he has over each department of service men competent and faithful to carry out in the most efficient manner their respective duties. It is in this last duty that he will find the most important field for his care and vigilance; and all his examination of details should be with reference to know how far he has been successful in this feature of his administration. Here lies the foundation of sound railway management. In the varied and numerous matters involved, it is idle for the chief executive officer to attempt much personal knowledge of details, and in devoting himself to these, except on occasion of special examination in reference to the duties of subordinates, or as they come incidentally, or by complaint to his notice, he will probably neglect his more appropriate and important duties.

It is not usual for the President to have an education which prepares him for his position, and he must be more or less embarrassed in many of his duties, particularly in making his selections of men for the charge of subordinate departments. He will be compelled to depend more or less on the recommendations of others, and these, even though well intended, are often much at fault, and lead to the appointment perhaps of respectable, at the same time unsuitable, men. Desirable as it is that the President should have a good knowledge of what is essential in every branch of the service he superintends, it is not often that this is secured, though high, and sometimes extravagant, compensation is paid to obtain it; and in this, as in other departments, a liberal salary is sure to obtain an incumbent, but not always the qualifications essential to the proper discharge of the duties required. The man, however, who is well fitted for the station, is cheap for the proprietors at any reasonable salary; and the unremitting cares that will press upon his attention, insure his bread from the sweat of his brow. His first duty will be to present to his Board for appointment as heads of the several departments, men who have education in the different branches of service to be provided for. His talent as a business man will be amply proved in the discharge of this duty. If by indiscretion—petty jealousy—superficial examination or partiality for personal friends, he makes bad selections, the proprietary interest must suffer; and on the other hand, if he proves himself a discreet man, with an eye single to the prosperity of the institution, having a frank and high-minded sense of duty, well experienced in the affairs he has assumed to manage, and no fear of rivals, rising from superior capacity, among his subordinates, he will gather around him a class of men that will produce order, regularity and efficiency through every department of the business of the institution. He may not succeed fully in the outset of his engagement, but he will ultimately produce the most favorable management.

Railways must generally be joint stock property, and from the impracticability of conducting the business in any other way, the power of management must be delegated to a portion of stock proprietors. A short railway involving a comparatively moderate amount of capital may be controlled by a small number of men, who may be residents in the vicinity, familiar with each other, and so situated as to have cognizance of much of its operations, and be able to discuss intelligently its doings and its business; and by personal attendance at annual meetings, investigate its management, and judge whether their influence and votes should be given to sustain, or to change the directors. Under these circumstances the railway is situated like much other joint stock property, where the limited character of operations renders it practicable to institute scrutiny into the administration of affairs. This will be the case with railways of moderate extent, and with the proprietors mostly residents on or near its line; but our most important lines are not often under so favorable cognizance, and in many cases the proprietors are so distant from the location and business of their property, and so little acquainted with each other, that they have little opportunity to know the details of their affairs, and scarcely pretend to any personal effort, beyond that of sending their proxy to such party in the management as may solicit it, with envelope stamped and directed for return at the least possible care and expense. It is not, as before observed, until dividends of profits fail, that proprietors begin to inquire into the management. If a single proprietor, or a number residing in the same vicinity, obtain the impression that their affairs are not well conducted, they seldom put forth any efficient proceedings for correction. Their comparatively small interest impresses on their minds the impotency of any effort they may make to produce a desired change of administration, and they usually submit to the loss, holding on, in the hope of better prospects, or selling out their interest at whatever the market will give.

If property in railways was generally held by men of higher wealth, and holding large interest in the property, there would be more practicability in efforts for any needed reform. This, however, is not generally the case; for the alluring prospects that have been held out by some very prosperous investments of this kind, on railways that, with a good traffic, have been well conducted, and by insnaring dividends paid by others, have induced many to invest their small means in this kind of property, into which the more wary capitalist has seldom intrusted his funds; and consequently the proprietors are numerous, and not often able to effect that concert of action so necessary to produce any desirable change. At first blush, it seems very proper and practicable for dissatisfied proprietors to make a change that appears necessary to secure their interest; but on reflection it will appear that a concentration sufficient in most cases to accomplish the object, will involve so much time, labor, and expense, that few will be found willing to undertake it; and as before remarked, the security of good management for this kind of property will, in most cases, be dependent on the election, in the first organization, of a competent and faithful Board of Directors, who will be disposed to give the attention required for the faithful and efficient conduct of the affairs of the institution. In the event of bad management, under the circumstances above alluded to, if the property does not go into the hands of the bond proprietors, it is probable the stock will fall so low that a few capitalists may combine to purchase it, or so much as may enable them to control it, and will have no difficulty in controlling the election and consequently the management; and if the railway has merits to induce such purchase, it will be likely to assume a new aspect, and may become a very useful institution for the trade of the district, and very remunerating to the new proprietors, and prove to be the best result for those of the old, who had not parted with their interest.

The fact is to be deplored, that men appointed to these high and honorable trusts—trusts involving the public welfare, and especially the interests of numerous confiding proprietors—should use the influence and power committed to their fidelity, to advance their personal interests and that of their personal friends. This has not arisen from the want of competent and faithful men in the community, for some institutions of this kind have been faithfully conducted, and the proprietors and the public have had little reason to complain. It has rather, and in the main, resulted from the opportunity afforded by a numerous constituency, for improper men to gain the control. That this is a difficulty in this class of property cannot be disputed, and no doubt it has been greatly aggravated by the zeal with which men of moderate means have sought this source of investment, under the impression that they would secure a large rate of income, with no trouble or care to manage the property. But it did not occur to them, that men of fair standing in the business world, might not have the virtue to resist temptation; and, perhaps with small beginnings, the unfaithful manager goes on from step to step, in a more or less rapid depletion, until the ruin of his property stares the hitherto confiding stockholder in the face. On the other hand, when this class of investers have fallen into good hands, they have, for the most part, according to the productiveness of the traffic of the line, been well cared for, and the income of their small property has been faithfully handed over, with small care or trouble on their part, and after years have rolled by, they find their property affording unimpaired income.

Notwithstanding these favorable instances, the railway business, in all its aspects, can hardly be said to be a judicious source of investment for small means, where the individual proprietor must be mainly dependent on his faith, for any views he may have of the interior of his affairs, and quite impotent to effect any correction in the event of mal-administration. Yet, strange as it may appear, of the large amount of capital that has gone into our railways, and especially for those that have turned out bad, a very great proportion has been furnished by small subscribers. The shrewd capitalist has wanted confidence in the strength of the traffic to afford adequate compensation, or he has had doubts as to the capacity or integrity of the men whom he sees in, or likely to be in, the control of affairs; and therefore holds back on the stock subscription, and waits for the issue of bonds. In the latter he is not so reserved—regarding the stock as adequate protection, he considers his security to be good; though the developments of a few years past have very much shaken the confidence that heretofore existed in this branch of security.

But whatever may have been the character of the capitalists that have hitherto furnished funds for the railways, there can be little doubt, those that have been broken down by mismanagement, as before observed, will eventually get into the hands of a comparatively small number of proprietors, who, by more faithful and efficient supervision, will be likely to make them subserve the great interest for which they were instituted. Railways of reasonable traffic, managed with skill and fidelity, and in some instances those for a time ill-managed, but having the mass of proprietors in near proximity, and able to coöperate effectively in promoting any desired change, will escape this result, and in some cases improve in productiveness. It cannot, however, be regarded safe for a stock proprietor to sit down and fold his hands; he must exercise the same care and watchfulness that is required in the successful management of all property, being cautious as to whose hands he commits his trust.

Our railways are, in the main, good property, though too much multiplied in some localities, and are worth what they have actually cost in cash, the depreciation is principally owing to a want of capacity and fidelity in their management, and where it may be practicable to correct this evil, they will for the most part be fairly remunerating to the proprietors, though the losses by the depletion through mal-administration may not be restored.

In view of the history of railway management, and of the vast importance of this kind of improvement, to the commercial, social and political interests of the country, it is humiliating, and deeply to be regretted, that so great a want of high minded fidelity should have been manifested in the admin istration of many of these works; so depleting the stock interest as in a great measure to destroy confidence in investments necessary to any further prosecution of enterprises so eminently characterizing the progress of our age.

Though it is small consolation to the stock proprietor, who has felt the loss of income, that could be ill spared as a means to meet his current necessities, the railways, even those badly managed, have very largely contributed to advance the facilities of intercourse, and to increase the general prosperity of the country; and this much beyond the loss that unprincipled and incompetent administration has wrested from the hard earnings and frugal savings of a great number of small, and of some large proprietors. We can hope for amendment, as before intimated, only in the improvement of our civilization, from which men will come to regard the honorable discharge of a trust as more to be esteemed and valued than any mass of ill-gotten and unjust gains, however the latter may be covered over by cunning craft. When the community come to look on the two classes, in the true light of their respective characters, railway property will be held with more confidence.

Have I been too severe on unfaithful management? I hope no upright managers have been wounded by anything I have said, for all of this class I highly appreciate, and have endeavored to keep this distinction clear. But from the nature of the trust—the necessity of confidence in councils, and transactions that cannot be seen—the great interest of the proprietors and the public in good and faithful managers—and the loose degree of morality that has been so developed in many cases, as manifestly to destroy confidence in a great degree, have demanded, that while the evil should be discussed dispassionately, it should be candidly and fully exposed. I do not feel conscious of having done any injustice, and have simply aimed to put proprietors on their guard, in reference to the management of their property, and the circumstances that tend to its destruction.

THE END.

CATALOGUE

OF

PRACTICAL AND SCIENTIFIC BOOKS,

PUBLISHED BY

HENRY CAREY BAIRD,

Industrial Publisher,

NO. 406 WALNUT STREET,

PHILADELPHIA.

☞ Any of the Books comprised in this Catalogue will be sent by mail, free of postage, at the publication price.

☞ A Descriptive Catalogue, 96 pages, 8vo., will be sent, free of postage, to any one who will furnish the publisher with his address.

ARLOT.—A Complete Guide for Coach Painters.

Translated from the French of M. ARLOT, Coach Painter; for eleven years Foreman of Painting to M. Eherler, Coach Maker, Paris. By A. A. FESQUET, Chemist and Engineer. To which is added an Appendix, containing Information respecting the Materials and the Practice of Coach and Car Painting and Varnishing in the United States and Great Britain. 12mo. $1.25

ARMENGAUD, AMOROUX, and JOHNSON.—The Practical Draughtsman's Book of Industrial Design, and Machinist's and Engineer's Drawing Companion:

Forming a Complete Course of Mechanical Engineering and Architectural Drawing. From the French of M. Armengaud the elder, Prof. of Design in the Conservatoire of Arts and Industry, Paris, and MM. Armengaud the younger, and Amoroux, Civil Engineers. Rewritten and arranged with additional matter and plates, selections from and examples of the most useful and generally employed mechanism of the day. By WILLIAM JOHNSON, Assoc. Inst. C. E., Editor of "The Practical Mechanic's Journal." Illustrated by 50 folio steel plates, and 50 wood-cuts. A new edition, 4to. $10.00

ARROWSMITH.—Paper-Hanger's Companion:

A Treatise in which the Practical Operations of the Trade are Systematically laid down: with Copious Directions Preparatory to Papering; Preventives against the Effect of Damp on Walls; the Various Cements and Pastes Adapted to the Several Purposes of the Trade; Observations and Directions for the Panelling and Ornamenting of Rooms, etc. By JAMES ARROWSMITH, Author of "Analysis of Drapery," etc. 12mo., cloth. $1.25

ASHTON.—The Theory and Practice of the Art of Designing Fancy Cotton and Woollen Cloths from Sample:

Giving full Instructions for Reducing Drafts, as well as the Methods of Spooling and Making out Harness for Cross Drafts, and Finding any Required Reed, with Calculations and Tables of Yarn. By FREDERICK T. ASHTON, Designer, West Pittsfield, Mass. With 52 Illustrations. One volume, 4to. $10.00

BAIRD.—Letters on the Crisis, the Currency and the Credit System.

By HENRY CAREY BAIRD. Pamphlet. 05

BAIRD.—Protection of Home Labor and Home Productions necessary to the Prosperity of the American Farmer.

By HENRY CAREY BAIRD. 8vo., paper. 10

BAIRD.—Some of the Fallacies of British Free-Trade Revenue Reform.

Two Letters to Arthur Latham Perry, Professor of History and Political Economy in Williams College. By HENRY CAREY BAIRD. Pamphlet. 05

BAIRD.—The Rights of American Producers, and the Wrongs of British Free-Trade Revenue Reform.

By HENRY CAREY BAIRD. Pamphlet. 05

BAIRD.—Standard Wages Computing Tables:

An Improvement in all former Methods of Computation, so arranged that wages for days, hours, or fractions of hours, at a specified rate per day or hour, may be ascertained at a glance. By T. SPANGLER BAIRD. Oblong folio. $5.00

BAIRD.—The American Cotton Spinner, and Manager's and Carder's Guide:

A Practical Treatise on Cotton Spinning; giving the Dimensions and Speed of Machinery, Draught and Twist Calculations, etc.; with notices of recent Improvements: together with Rules and Examples for making changes in the sizes and numbers of Roving and Yarn. Compiled from the papers of the late ROBERT H. BAIRD. 12mo. $1.50

BAKER.—Long-Span Railway Bridges :

Comprising Investigations of the Comparative Theoretical and Practical Advantages of the various Adopted or Proposed Type Systems of Construction; with numerous Formulæ and Tables. By B. BAKER. 12mo. $2.00

BAUERMAN.—A Treatise on the Metallurgy of Iron :

Containing Outlines of the History of Iron Manufacture, Methods of Assay, and Analysis of Iron Ores, Processes of Manufacture of Iron and Steel, etc., etc. By H. BAUERMAN, F. G. S., Associate of the Royal School of Mines. First American Edition, Revised and Enlarged. With an Appendix on the Martin Process for Making Steel, from the Report of ABRAM S. HEWITT, U. S. Commissioner to the Universal Exposition at Paris, 1867. Illustrated. 12mo. . $2.00

BEANS.—A Treatise on Railway Curves and the Location of Railways.

By E. W. BEANS, C. E. Illustrated. 12mo. Tucks. . . $1.50

BELL.—Carpentry Made Easy :

Or, The Science and Art of Framing on a New and Improved System. With Specific Instructions for Building Balloon Frames, Barn Frames, Mill Frames, Warehouses, Church Spires, etc. Comprising also a System of Bridge Building, with Bills, Estimates of Cost, and valuable Tables. Illustrated by 38 plates, comprising nearly 200 figures. By WILLIAM E. BELL, Architect and Practical Builder. 8vo. . $5.00

BELL.—Chemical Phenomena of Iron Smelting :

An Experimental and Practical Examination of the Circumstances which determine the Capacity of the Blast Furnace, the Temperature of the Air, and the proper Condition of the Materials to be operated upon. By I. LOWTHIAN BELL. Illustrated. 8vo. . . $6.00

BEMROSE.—Manual of Wood Carving :

With Practical Illustrations for Learners of the Art, and Original and Selected Designs. By WILLIAM BEMROSE, Jr. With an Introduction by LLEWELLYN JEWITT, F. S. A., etc. With 128 Illustrations. 4to., cloth. $3.00

BICKNELL.—Village Builder, and Supplement :

Elevations and Plans for Cottages, Villas, Suburban Residences, Farm Houses, Stables and Carriage Houses, Store Fronts, School Houses, Churches, Court Houses, and a model Jail; also, Exterior and Interior details for Public and Private Buildings, with approved Forms of Contracts and Specifications, including Prices of Building Materials and Labor at Boston, Mass., and St. Louis, Mo. Containing 75 plates drawn to scale; showing the style and cost of building in different sections of the country, being an original work comprising the designs of twenty leading architects, representing the New England, Middle, Western, and Southwestern States. 4to. . $12.00

BLENKARN.—Practical Specifications of Works executed in Architecture, Civil and Mechanical Engineering, and in Road Making and Sewering:

To which are added a series of practically useful Agreements and Reports. By JOHN BLENKARN. Illustrated by 15 large folding plates. 8vo. $9.00

BLINN.—A Practical Workshop Companion for Tin, Sheet-Iron, and Copperplate Workers:

Containing Rules for describing various kinds of Patterns used by Tin, Sheet-Iron, and Copper-plate Workers; Practical Geometry; Mensuration of Surfaces and Solids; Tables of the Weights of Metals, Lead Pipe, etc.; Tables of Areas and Circumferences of Circles; Japan, Varnishes, Lackers, Cements, Compositions, etc., etc. By LEROY J. BLINN, Master Mechanic. With over 100 Illustrations. 12mo. $2.50

BOOTH.—Marble Worker's Manual:

Containing Practical Information respecting Marbles in general, their Cutting, Working, and Polishing; Veneering of Marble; Mosaics; Composition and Use of Artificial Marble, Stuccos, Cements, Receipts, Secrets, etc., etc. Translated from the French by M. L. BOOTH. With an Appendix concerning American Marbles. 12mo., cloth. $1.50

BOOTH AND MORFIT.—The Encyclopedia of Chemistry, Practical and Theoretical:

Embracing its application to the Arts, Metallurgy, Mineralogy, Geology, Medicine, and Pharmacy. By JAMES C. BOOTH, Melter and Refiner in the United States Mint, Professor of Applied Chemistry in the Franklin Institute, etc., assisted by CAMPBELL MORFIT, author of "Chemical Manipulations," etc. Seventh edition. Royal 8vo., 978 pages, with numerous wood-cuts and other illustrations. . $5.00

BOX.—A Practical Treatise on Heat:

As applied to the Useful Arts; for the Use of Engineers, Architects, etc. By THOMAS BOX, author of "Practical Hydraulics." Illustrated by 14 plates containing 114 figures. 12mo. $4.25

BOX.—Practical Hydraulics:

A Series of Rules and Tables for the use of Engineers, etc. By THOMAS BOX. 12mo. $2.50

BROWN.—Five Hundred and Seven Mechanical Movements:

Embracing all those which are most important in Dynamics, Hydraulics, Hydrostatics, Pneumatics, Steam Engines, Mill and other Gearing, Presses, Horology, and Miscellaneous Machinery; and including many movements never before published, and several of which have only recently come into use. By HENRY T. BROWN, Editor of the "American Artisan." In one volume, 12mo. . . . $1.00

BUCKMASTER.—The Elements of Mechanical Physics:

By J. C. BUCKMASTER, late Student in the Government School of Mines; Certified Teacher of Science by the Department of Science and Art; Examiner in Chemistry and Physics in the Royal College of Preceptors; and late Lecturer in Chemistry and Physics of the Royal Polytechnic Institute. Illustrated with numerous engravings. In one volume, 12mo. $1.50

BULLOCK.—The American Cottage Builder:

A Series of Designs, Plans, and Specifications, from $200 to $20,000, for Homes for the People; together with Warming, Ventilation, Drainage, Painting, and Landscape Gardening. By JOHN BULLOCK, Architect, Civil Engineer, Mechanician, and Editor of "The Rudiments of Architecture and Building," etc., etc. Illustrated by 75 engravings. In one volume, 8vo. $3.50

BULLOCK.—The Rudiments of Architecture and Building:

For the use of Architects, Builders, Draughtsmen, Machinists, Engineers, and Mechanics. Edited by JOHN BULLOCK, author of "The American Cottage Builder." Illustrated by 250 engravings. In one volume, 8vo. $3.50

BURGH.—Practical Illustrations of Land and Marine Engines:

Showing in detail the Modern Improvements of High and Low Pressure, Surface Condensation, and Super-heating, together with Land and Marine Boilers. By N. P. BURGH, Engineer. Illustrated by 20 plates, double elephant folio, with text. . . . $21.00

BURGH.—Practical Rules for the Proportions of Modern Engines and Boilers for Land and Marine Purposes.

By N. P. BURGH, Engineer. 12mo. $1.50

BURGH.—The Slide-Valve Practically Considered.

By N. P. BURGH, Engineer. Completely illustrated. 12mo. $2.00

BYLES.—Sophisms of Free Trade and Popular Political Economy Examined.

By a BARRISTER (Sir JOHN BARNARD BYLES, Judge of Common Pleas). First American from the Ninth English Edition, as published by the Manchester Reciprocity Association. In one volume, 12mo. Paper, 75 cts. Cloth $1.25

BYRN.—The Complete Practical Brewer:

Or Plain, Accurate, and Thorough Instructions in the Art of Brewing Beer, Ale, Porter, including the Process of making Bavarian Beer, all the Small Beers, such as Root-beer, Ginger-pop, Sarsaparilla-beer, Mead, Spruce Beer, etc., etc. Adapted to the use of Public Brewers and Private Families. By M. LA FAYETTE BYRN, M. D. With illustrations. 12mo. $1.25

BYRN.—The Complete Practical Distiller:

Comprising the most perfect and exact Theoretical and Practical Description of the Art of Distillation and Rectification; including all of the most recent improvements in distilling apparatus; instructions for preparing spirits from the numerous vegetables, fruits, etc.; directions for the distillation and preparation of all kinds of brandies and other spirits, spirituous and other compounds, etc., etc. By M. LA FAYETTE BYRN, M. D. Eighth Edition. To which are added, Practical Directions for Distilling, from the French of Th. Fling, Brewer and Distiller. 12mo. $1.50

BYRNE.—Handbook for the Artisan, Mechanic, and Engineer:

Comprising the Grinding and Sharpening of Cutting Tools, Abrasive Processes, Lapidary Work, Gem and Glass Engraving, Varnishing and Lackering, Apparatus, Materials and Processes for Grinding and Polishing, etc. By OLIVER BYRNE. Illustrated by 185 wood engravings. In one volume, 8vo. $5.00

BYRNE.—Pocket Book for Railroad and Civil Engineers:

Containing New, Exact, and Concise Methods for Laying out Railroad Curves, Switches, Frog Angles, and Crossings; the Staking out of work; Levelling; the Calculation of Cuttings; Embankments; Earth-work, etc. By OLIVER BYRNE. 18mo., full bound, pocket-book form $1.75

BYRNE.—The Practical Model Calculator:

For the Engineer, Mechanic, Manufacturer of Engine Work, Naval Architect, Miner, and Millwright. By OLIVER BYRNE. 1 volume, 8vo., nearly 600 pages $4.50

BYRNE.—The Practical Metal-Worker's Assistant:

Comprising Metallurgic Chemistry; the Arts of Working all Metals and Alloys; Forging of Iron and Steel; Hardening and Tempering; Melting and Mixing; Casting and Founding; Works in Sheet Metal; The Processes Dependent on the Ductility of the Metals; Soldering; and the most Improved Processes and Tools employed by Metal-Workers. With the Application of the Art of Electro-Metallurgy to Manufacturing Processes; collected from Original Sources, and from the Works of Holtzapffel, Bergeron, Leupold, Plumier, Napier, Scoffern, Clay, Fairbairn, and others. By OLIVER BYRNE. A new, revised, and improved edition, to which is added An Appendix, containing THE MANUFACTURE OF RUSSIAN SHEET-IRON. By JOHN PERCY, M. D., F.R.S. THE MANUFACTURE OF MALLEABLE IRON CASTINGS, and IMPROVEMENTS IN BESSEMER STEEL. By A. A. FESQUET, Chemist and Engineer. With over 600 Engravings, illustrating every Branch of the Subject. 8vo. $7.00

Cabinet Maker's Album of Furniture:

Comprising a Collection of Designs for Furniture. Illustrated by 48 Large and Beautifully Engraved Plates. In one vol., oblong $5.00

CALLINGHAM.—Sign Writing and Glass Embossing:

A Complete Practical Illustrated Manual of the Art. By JAMES CALLINGHAM. In one volume, 12mo. $1.50

CAMPIN.—A Practical Treatise on Mechanical Engineering:

Comprising Metallurgy, Moulding, Casting, Forging, Tools, Workshop Machinery, Mechanical Manipulation, Manufacture of Steam-engines, etc., etc. With an Appendix on the Analysis of Iron and Iron Ores. By FRANCIS CAMPIN, C. E. To which are added, Observations on the Construction of Steam Boilers, and Remarks upon Furnaces used for Smoke Prevention; with a Chapter on Explosions. By R. Armstrong, C. E., and John Bourne. Rules for Calculating the Change Wheels for Screws on a Turning Lathe, and for a Wheel-cutting Machine. By J. LA NICCA. Management of Steel, Including Forging, Hardening, Tempering, Annealing, Shrinking, and Expansion. And the Case-hardening of Iron. By G. EDE. 8vo. Illustrated with 29 plates and 100 wood engravings . . . $6.00

CAMPIN.—The Practice of Hand-Turning in Wood, Ivory, Shell, etc.:

With Instructions for Turning such works in Metal as may be required in the Practice of Turning Wood, Ivory, etc. Also, an Appendix on Ornamental Turning. By FRANCIS CAMPIN; with Numerous Illustrations. 12mo., cloth $3.00

CAREY.—The Works of Henry C. Carey:

FINANCIAL CRISES, their Causes and Effects. 8vo. paper . 25

HARMONY OF INTERESTS: Agricultural, Manufacturing, and Commercial. 8vo., cloth $1.50

MANUAL OF SOCIAL SCIENCE. Condensed from Carey's "Principles of Social Science." By KATE MCKEAN. 1 vol. 12mo. $2.25

MISCELLANEOUS WORKS: comprising "Harmony of Interests," "Money," "Letters to the President," "Financial Crises," "The Way to Outdo England Without Fighting Her," "Resources of the Union," "The Public Debt," "Contraction or Expansion?" "Review of the Decade 1857-'67," "Reconstruction," etc., etc. Two vols., 8vo., cloth $10.00

PAST, PRESENT, AND FUTURE. 8vo. $2.50

PRINCIPLES OF SOCIAL SCIENCE. 3 vols., 8vo., cloth $10.00

THE SLAVE-TRADE, DOMESTIC AND FOREIGN; Why it Exists, and How it may be Extinguished (1853). 8vo., cloth . $2.00

LETTERS ON INTERNATIONAL COPYRIGHT (1867) . 50

THE UNITY OF LAW: As Exhibited in the Relations of Physical, Social, Mental, and Moral Science (1872). In one volume, 8vo., pp. xxiii., 433. Cloth $3.50

CHAPMAN.—A Treatise on Ropemaking:

As Practised in private and public Rope yards, with a Description of the Manufacture, Rules, Tables of Weights, etc., adapted to the Trades, Shipping, Mining, Railways, Builders, etc. By ROBERT CHAPMAN. 24mo. $1.50

COLBURN.—The Locomotive Engine:

Including a Description of its Structure, Rules for Estimating its Capabilities, and Practical Observations on its Construction and Management. By ZERAH COLBURN. Illustrated. A new edition. 12mo. $1.25

CRAIK.—The Practical American Millwright and Miller.

By DAVID CRAIK, Millwright. Illustrated by numerous wood engravings, and two folding plates. 8vo. $5.00

DE GRAFF.—The Geometrical Stair Builders' Guide:

Being a Plain Practical System of Hand-Railing, embracing all its necessary Details, and Geometrically Illustrated by 22 Steel Engravings; together with the use of the most approved principles of Practical Geometry. By SIMON DE GRAFF, Architect. 4to. . $5.00

DE KONINCK.—DIETZ.—A Practical Manual of Chemical Analysis and Assaying:

As applied to the Manufacture of Iron from its Ores, and to Cast Iron, Wrought Iron, and Steel, as found in Commerce. By L. L. DE KONINCK, Dr. Sc., and E. DIETZ, Engineer. Edited with Notes, by ROBERT MALLET, F.R.S., F.S.G., M.I.C.E., etc. American Edition, Edited with Notes and an Appendix on Iron Ores, by A. A. FESQUET, Chemist and Engineer. One volume, 12mo. $2.50

DUNCAN.—Practical Surveyor's Guide:

Containing the necessary information to make any person, of common capacity, a finished land surveyor without the aid of a teacher. By ANDREW DUNCAN. Illustrated. 12mo., cloth. . . . $1.25

DUPLAIS.—A Treatise on the Manufacture and Distillation of Alcoholic Liquors:

Comprising Accurate and Complete Details in Regard to Alcohol from Wine, Molasses, Beets, Grain, Rice, Potatoes, Sorghum, Asphodel, Fruits, etc.; with the Distillation and Rectification of Brandy, Whiskey, Rum, Gin, Swiss Absinthe, etc., the Preparation of Aromatic Waters, Volatile Oils or Essences, Sugars, Syrups, Aromatic Tinctures, Liqueurs, Cordial Wines, Effervescing Wines, etc., the Aging of Brandy and the Improvement of Spirits, with Copious Directions and Tables for Testing and Reducing Spirituous Liquors, etc., etc. Translated and Edited from the French of MM. DUPLAIS, Ainé et Jeune. By M. McKENNIE, M.D. To which are added the United States Internal Revenue Regulations for the Assessment and Collection of Taxes on Distilled Spirits. Illustrated by fourteen folding plates and several wood engravings. 743 pp., 8vo. $10.00

DUSSAUCE.—A General Treatise on the Manufacture of Every Description of Soap:

Comprising the Chemistry of the Art, with Remarks on Alkalies, Saponifiable Fatty Bodies, the apparatus necessary in a Soap Factory, Practical Instructions in the manufacture of the various kinds of Soap, the assay of Soaps, etc., etc. Edited from Notes of Larmé, Fontenelle, Malapayre, Dufour, and others, with large and important additions by Prof. H. DUSSAUCE, Chemist. Illustrated. In one vol., 8vo. . $10.00

DUSSAUCE.—A General Treatise on the Manufacture of Vinegar:

Theoretical and Practical. Comprising the various Methods, by the Slow and the Quick Processes, with Alcohol, Wine, Grain, Malt, Cider, Molasses, and Beets; as well as the Fabrication of Wood Vinegar, etc., etc. By Prof. H. DUSSAUCE. In one volume, 8vo. . . $5.00

DUSSAUCE.—A New and Complete Treatise on the Arts of Tanning, Currying, and Leather Dressing:

Comprising all the Discoveries and Improvements made in France, Great Britain, and the United States. Edited from Notes and Documents of Messrs. Sallerou, Grouvelle, Duval, Dessables, Labarraque, Payen, René, De Fontenelle, Malapeyre, etc., etc. By Prof. H. DUSSAUCE, Chemist. Illustrated by 212 wood engravings. 8vo. $10.00

DUSSAUCE.—A Practical Guide for the Perfumer:

Being a New Treatise on Perfumery, the most favorable to the Beauty without being injurious to the Health, comprising a Description of the substances used in Perfumery, the Formulæ of more than 1000 Preparations, such as Cosmetics, Perfumed Oils, Tooth Powders, Waters, Extracts, Tinctures, Infusions, Spirits, Vinaigres, Essential Oils, Pastels, Creams, Soaps, and many new Hygienic Products not hitherto described. Edited from Notes and Documents of Messrs. Debay, Lanel, etc. With additions by Prof. H. DUSSAUCE, Chemist. 12mo. $3.00

DUSSAUCE.—Practical Treatise on the Fabrication of Matches, Gun Cotton, and Fulminating Powders.

By Prof. H. DUSSAUCE. 12mo. $3.00

Dyer and Color-maker's Companion:

Containing upwards of 200 Receipts for making Colors, on the most approved principles, for all the various styles and fabrics now in existence; with the Scouring Process, and plain Directions for Preparing, Washing-off, and Finishing the Goods. In one vol., 12mo. . $1.25

EASTON.—A Practical Treatise on Street or Horse-power Railways.

By ALEXANDER EASTON, C.E. Illustrated by 23 plates. 8vo., cloth. $2.00

ELDER.—Questions of the Day:

Economic and Social. By Dr. WILLIAM ELDER. 8vo. . $3.00

FAIRBAIRN.—The Principles of Mechanism and Machinery of Transmission:

Comprising the Principles of Mechanism, Wheels, and Pulleys, Strength and Proportions of Shafts, Coupling of Shafts, and Engaging and Disengaging Gear. By Sir WILLIAM FAIRBAIRN, C.E., LL.D., F.R.S., F.G.S. Beautifully illustrated by over 150 wood-cuts. In one volume, 12mo. $2.50

FORSYTH.—Book of Designs for Headstones, Mural, and other Monuments:

Containing 78 Designs. By JAMES FORSYTH. With an Introduction by CHARLES BOUTELL, M. A. 4to., cloth. $5.00

GIBSON.—The American Dyer:

A Practical Treatise on the Coloring of Wool, Cotton, Yarn and Cloth, in three parts. Part First gives a descriptive account of the Dye Stuffs; if of vegetable origin, where produced, how cultivated, and how prepared for use; if chemical, their composition, specific gravities, and general adaptability, how adulterated, and how to detect the adulterations, etc. Part Second is devoted to the Coloring of Wool, giving recipes for one hundred and twenty-nine different colors or shades, and is supplied with sixty colored samples of Wool. Part Third is devoted to the Coloring of Raw Cotton or Cotton Waste, for mixing with Wool Colors in the Manufacture of all kinds of Fabrics, gives recipes for thirty-eight different colors or shades, and is supplied with twenty-four colored samples of Cotton Waste. Also, recipes for Coloring Beavers, Doeskins, and Flannels, with remarks upon Anilines, giving recipes for fifteen different colors or shades, and nine samples of Aniline Colors that will stand both the Fulling and Scouring process. Also, recipes for Aniline Colors on Cotton Thread, and recipes for Common Colors on Cotton Yarns. Embracing in all over two hundred recipes for Colors and Shades, and ninety-four samples of Colored Wool and Cotton Waste, etc. By RICHARD H. GIBSON, Practical Dyer and Chemist. In one volume, 8vo. . . $12.50

GILBART.—History and Principles of Banking:

A Practical Treatise. By JAMES W. GILBART, late Manager of the London and Westminster Bank. With additions. In one volume, 8vo., 600 pages, sheep $5.00

Gothic Album for Cabinet Makers:

Comprising a Collection of Designs for Gothic Furniture. Illustrated by 23 large and beautifully engraved plates. Oblong . . $3.00

GRANT.—Beet-root Sugar and Cultivation of the Beet.

By E. B. GRANT. 12mo. $1.25

GREGORY.—Mathematics for Practical Men:

Adapted to the Pursuits of Surveyors, Architects, Mechanics, and Civil Engineers. By OLINTHUS GREGORY. 8vo., plates, cloth $3.00

GRISWOLD.—Railroad Engineer's Pocket Companion for the Field:

Comprising Rules for Calculating Deflection Distances and Angles, Tangential Distances and Angles, and all Necessary Tables for Engineers; also the art of Levelling from Preliminary Survey to the Construction of Railroads, intended Expressly for the Young Engineer, together with Numerous Valuable Rules and Examples. By W. GRISWOLD. 12mo., tucks $1.75

GRUNER.—Studies of Blast Furnace Phenomena.

By M. L. GRUNER, President of the General Council of Mines of France, and lately Professor of Metallurgy at the Ecole des Mines. Translated, with the Author's sanction, with an Appendix, by L. D. B. Gordon, F. R. S. E., F. G. S. Illustrated. 8vo. $2.50

GUETTIER.—Metallic Alloys:

Being a Practical Guide to their Chemical and Physical Properties, their Preparation, Composition, and Uses. Translated from the French of A. GUETTIER, Engineer and Director of Foundries, author of "La Fouderie en France," etc., etc. By A. A. FESQUET, Chemist and Engineer. In one volume, 12mo. $3.00

HARRIS.—Gas Superintendent's Pocket Companion.

By HARRIS & BROTHER, Gas Meter Manufacturers, 1115 and 1117 Cherry Street, Philadelphia. Full bound in pocket-book form $2.00

Hats and Felting:

A Practical Treatise on their Manufacture. By a Practical Hatter. Illustrated by Drawings of Machinery, etc. 8vo. . . . $1.25

HOFMANN.—A Practical Treatise on the Manufacture of Paper in all its Branches.

By CARL HOFMANN. Late Superintendent of paper mills in Germany and the United States; recently manager of the Public Ledger Paper Mills, near Elkton, Md. Illustrated by 110 wood engravings, and five large folding plates. In one volume, 4to., cloth; 398 pages $15.00

HUGHES.—American Miller and Millwright's Assistant.

By WM. CARTER HUGHES. A new edition. In one vol., 12mo. $1.50

HURST.—A Hand-Book for Architectural Surveyors and others engaged in Building:

Containing Formulæ useful in Designing Builder's work, Table of Weights, of the materials used in Building, Memoranda connected with Builders' work, Mensuration, the Practice of Builders' Measurement, Contracts of Labor, Valuation of Property, Summary of the Practice in Dilapidation, etc., etc. By J. F. HURST, C. E. Second edition, pocket-book form, full bound $2.50

JERVIS.—Railway Property:

A Treatise on the Construction and Management of Railways; designed to afford useful knowledge, in the popular style, to the holders of this class of property; as well as Railway Managers, Officers, and Agents. By JOHN B. JERVIS, late Chief Engineer of the Hudson River Railroad, Croton Aqueduct, etc. In one vol., 12mo., cloth $2.00

JOHNSTON.—Instructions for the Analysis of Soils, Limestones, and Manures.

By J. F. W. JOHNSTON. 12mo. 38

KEENE.—A Hand-Book of Practical Gauging:

For the Use of Beginners, to which is added, A Chapter on Distillation, describing the process in operation at the Custom House for ascertaining the strength of wines. By JAMES B. KEENE, of H. M. Customs. 8vo. $1.25

KELLEY.—Speeches, Addresses, and Letters on Industrial and Financial Questions.

By Hon. WILLIAM D. KELLEY, M. C. In one volume, 544 pages, 8vo. $3.00

KENTISH.—A Treatise on a Box of Instruments,

And the Slide Rule; with the Theory of Trigonometry and Logarithms, including Practical Geometry, Surveying, Measuring of Timber, Cask and Malt Gauging, Heights, and Distances. By THOMAS KENTISH. In one volume. 12mo. $1.25

KOBELL.—ERNI.—Mineralogy Simplified:

A short Method of Determining and Classifying Minerals, by means of simple Chemical Experiments in the Wet Way. Translated from the last German Edition of F. VON KOBELL, with an Introduction to Blow-pipe Analysis and other additions. By HENRI ERNI, M. D., late Chief Chemist, Department of Agriculture, author of "Coal Oil and Petroleum." In one volume, 12mo. $2.50

LANDRIN.—A Treatise on Steel:

Comprising its Theory, Metallurgy, Properties, Practical Working, and Use. By M. H. C. LANDRIN, Jr., Civil Engineer. Translated from the French, with Notes, by A. A. FESQUET, Chemist and Engineer. With an Appendix on the Bessemer and the Martin Processes for Manufacturing Steel, from the Report of Abram S. Hewitt, United States Commissioner to the Universal Exposition, Paris, 1867. In one volume, 12mo. $3.00

LARKIN.—The Practical Brass and Iron Founder's Guide:

A Concise Treatise on Brass Founding, Moulding, the Metals and their Alloys, etc.: to which are added Recent Improvements in the Manufacture of Iron, Steel by the Bessemer Process, etc., etc. By JAMES LARKIN, late Conductor of the Brass Foundry Department in Reany, Neafie & Co's. Penn Works, Philadelphia. Fifth edition, revised, with Extensive additions. In one volume, 12mo. . . . $2.25

LEAVITT.—Facts about Peat as an Article of Fuel:

With Remarks upon its Origin and Composition, the Localities in which it is found, the Methods of Preparation and Manufacture, and the various Uses to which it is applicable; together with many other matters of Practical and Scientific Interest. To which is added a chapter on the Utilization of Coal Dust with Peat for the Production of an Excellent Fuel at Moderate Cost, specially adapted for Steam Service. By T. H. LEAVITT. Third edition. 12mo. $1.75

LEROUX, C.—A Practical Treatise on the Manufacture of Worsteds and Carded Yarns:

Comprising Practical Mechanics, with Rules and Calculations applied to Spinning; Sorting, Cleaning, and Scouring Wools; the English and French methods of Combing, Drawing, and Spinning Worsteds and Manufacturing Carded Yarns. Translated from the French of CHARLES LEROUX, Mechanical Engineer, and Superintendent of a Spinning Mill, by HORATIO PAINE, M. D., and A. A. FESQUET, Chemist and Engineer. Illustrated by 12 large Plates. To which is added an Appendix, containing extracts from the Reports of the International Jury, and of the Artisans selected by the Committee appointed by the Council of the Society of Arts, London, on Woollen and Worsted Machinery and Fabrics, as exhibited in the Paris Universal Exposition, 1867. 8vo., cloth. $5.00

LESLIE (Miss).—Complete Cookery:

Directions for Cookery in its Various Branches. By MISS LESLIE. 60th thousand. Thoroughly revised, with the addition of New Receipts. In one volume, 12mo., cloth. $1.50

LESLIE (Miss).—Ladies' House Book:

A Manual of Domestic Economy. 20th revised edition. 12mo., cloth.

LESLIE (Miss).—Two Hundred Receipts in French Cookery.

Cloth, 12mo.

LIEBER.—Assayer's Guide:

Or, Practical Directions to Assayers, Miners, and Smelters, for the Tests and Assays, by Heat and by Wet Processes, for the Ores of all the principal Metals, of Gold and Silver Coins and Alloys, and of Coal, etc. By OSCAR M. LIEBER. 12mo., cloth. . . $1.25

LOTH.—The Practical Stair Builder:

A Complete Treatise on the Art of Building Stairs and Hand-Rails, Designed for Carpenters, Builders, and Stair-Builders. Illustrated with Thirty Original Plates. By C. EDWARD LOTH, Professional Stair-Builder. One large 4to. volume. $10.00

LOVE.—The Art of Dyeing, Cleaning, Scouring, and Finishing, on the Most Approved English and French Methods:

Being Practical Instructions in Dyeing Silks, Woollens, and Cottons, Feathers, Chips, Straw, etc. Scouring and Cleaning Bed and Window Curtains, Carpets, Rugs, etc. French and English Cleaning, any Color or Fabric of Silk, Satin, or Damask. By THOMAS LOVE, a Working Dyer and Scourer. Second American Edition, to which are added General Instructions for the Use of Aniline Colors. In one volume, 8vo., 343 pages. $5.00

MAIN and BROWN.—Questions on Subjects Connected with the Marine Steam-Engine:

And Examination Papers: with Hints for their Solution. By THOMAS J. MAIN, Professor of Mathematics, Royal Naval College, and THOMAS BROWN, Chief Engineer, R. N. 12mo., cloth. . . . $1.50

MAIN and BROWN.—The Indicator and Dynamometer:

With their Practical Applications to the Steam-Engine. By THOMAS J. MAIN, M. A. F. R., Assistant Professor Royal Naval College, Portsmouth, and THOMAS BROWN, Assoc. Inst. C. E., Chief Engineer, R. N., attached to the Royal Naval College. Illustrated. From the Fourth London Edition. 8vo. $1.50

MAIN and BROWN.—The Marine Steam-Engine.

By THOMAS J. MAIN, F. R.; Assistant S. Mathematical Professor at the Royal Naval College, Portsmouth, and THOMAS BROWN, Assoc. Inst. C. E., Chief Engineer R. N. Attached to the Royal Naval College. Authors of "Questions connected with the Marine Steam-Engine," and the "Indicator and Dynamometer." With numerous Illustrations. In one volume, 8vo. $5.00

MARTIN.—Screw-Cutting Tables, for the Use of Mechanical Engineers:

Showing the Proper Arrangement of Wheels for Cutting the Threads of Screws of any required Pitch; with a Table for Making the Universal Gas-Pipe Thread and Taps. By W. A. MARTIN, Engineer. 8vo. 50

Mechanics' (Amateur) Workshop:

A treatise containing plain and concise directions for the manipulation of Wood and Metals, including Casting, Forging, Brazing, Soldering, and Carpentry. By the author of the "Lathe and its Uses." Third edition. Illustrated. 8vo. $3.00

MOLESWORTH.—Pocket-Book of Useful Formulæ and Memoranda for Civil and Mechanical Engineers.

By GUILFORD L. MOLESWORTH, Member of the Institution of Civil Engineers, Chief Resident Engineer of the Ceylon Railway. Second American, from the Tenth London Edition. In one volume, full bound in pocket-book form. $2.00

NAPIER.—A System of Chemistry Applied to Dyeing.

By JAMES NAPIER, F. C. S. A New and Thoroughly Revised Edition. Completely brought up to the present state of the Science, including the Chemistry of Coal Tar Colors, by A. A. FESQUET, Chemist and Engineer. With an Appendix on Dyeing and Calico Printing, as shown at the Universal Exposition, Paris, 1867. Illustrated. In one volume, 8vo., 422 pages. $5.00

NAPIER.—Manual of Electro-Metallurgy:

Including the Application of the Art to Manufacturing Processes. By JAMES NAPIER. Fourth American, from the Fourth London edition, revised and enlarged. Illustrated by engravings. In one vol., 8vo. $2.00

NASON.—Table of Reactions for Qualitative Chemical Analysis.

By HENRY B. NASON, Professor of Chemistry in the Rensselaer Polytechnic Institute, Troy, New York. Illustrated by Colors. . 63

NEWBERY.—Gleanings from Ornamental Art of every style:

Drawn from Examples in the British, South Kensington, Indian, Crystal Palace, and other Museums, the Exhibitions of 1851 and 1862, and the best English and Foreign works. In a series of one hundred exquisitely drawn Plates, containing many hundred examples. By ROBERT NEWBERY. 4to. $15.00

NICHOLSON.—A Manual of the Art of Bookbinding:

Containing full instructions in the different Branches of Forwarding, Gilding, and Finishing. Also, the Art of Marbling Book-edges and Paper. By JAMES B. NICHOLSON. Illustrated. 12mo., cloth. $2.25

NICHOLSON.—The Carpenter's New Guide:

A Complete Book of Lines for Carpenters and Joiners. By PETER NICHOLSON. The whole carefully and thoroughly revised by H. K. DAVIS, and containing numerous new and improved and original Designs for Roofs, Domes, etc. By SAMUEL SLOAN, Architect. Illustrated by 80 plates. 4to. $4.50

NORRIS.—A Hand-book for Locomotive Engineers and Machinists:

Comprising the Proportions and Calculations for Constructing Locomotives; Manner of Setting Valves; Tables of Squares, Cubes, Areas, etc., etc. By SEPTIMUS NORRIS, Civil and Mechanical Engineer. New edition. Illustrated. 12mo., cloth. $2.00

NYSTROM.—On Technological Education, and the Construction of Ships and Screw Propellers:

For Naval and Marine Engineers. By JOHN W. NYSTROM, late Acting Chief Engineer, U. S. N. Second edition, revised with additional matter. Illustrated by seven engravings. 12mo. . . $1.50

O'NEILL.—A Dictionary of Dyeing and Calico Printing:

Containing a brief account of all the Substances and Processes in use in the Art of Dyeing and Printing Textile Fabrics; with Practical Receipts and Scientific Information. By CHARLES O'NEILL, Analytical Chemist; Fellow of the Chemical Society of London; Member of the Literary and Philosophical Society of Manchester; Author of "Chemistry of Calico Printing and Dyeing." To which is added an Essay on Coal Tar Colors and their application to Dyeing and Calico Printing. By A. A. FESQUET, Chemist and Engineer. With an Appendix on Dyeing and Calico Printing, as shown at the Universal Exposition, Paris, 1867. In one volume, 8vo., 491 pages. . $6.00

ORTON.—Underground Treasures:

How and Where to Find Them. A Key for the Ready Determination of all the Useful Minerals within the United States. By JAMES ORTON, A. M Illustrated, 12mo. $1.50

OSBORN.—American Mines and Mining:

Theoretically and Practically Considered. By Prof. H. S. OSBORN. Illustrated by numerous engravings. 8vo. (*In preparation.*)

OSBORN.—The Metallurgy of Iron and Steel:

Theoretical and Practical in all its Branches; with special reference to American Materials and Processes. By H. S. OSBORN, LL. D., Professor of Mining and Metallurgy in Lafayette College, Easton, Pennsylvania. Illustrated by numerous large folding plates and wood-engravings. 8vo. $15.00

OVERMAN.—The Manufacture of Steel:

Containing the Practice and Principles of Working and Making Steel. A Handbook for Blacksmiths and Workers in Steel and Iron, Wagon Makers, Die Sinkers, Cutlers, and Manufacturers of Files and Hardware, of Steel and Iron, and for Men of Science and Art. By FREDERICK OVERMAN, Mining Engineer, Author of the "Manufacture of Iron," etc. A new, enlarged, and revised Edition. By A. A. FESQUET, Chemist and Engineer. $1.50

OVERMAN.—The Moulder and Founder's Pocket Guide:

A Treatise on Moulding and Founding in Green-sand, Dry-sand, Loam, and Cement; the Moulding of Machine Frames, Mill-gear, Hollow-ware, Ornaments, Trinkets, Bells, and Statues; Description of Moulds for Iron, Bronze, Brass, and other Metals; Plaster of Paris, Sulphur, Wax, and other articles commonly used in Casting; the Construction of Melting Furnaces, the Melting and Founding of Metals; the Composition of Alloys and their Nature. With an Appendix containing Receipts for Alloys, Bronze, Varnishes and Colors for Castings; also, Tables on the Strength and other qualities of Cast Metals. By FREDERICK OVERMAN, Mining Engineer, Author of "The Manufacture of Iron." With 42 Illustrations. 12mo. $1.50

Painter, Gilder, and Varnisher's Companion:

Containing Rules and Regulations in everything relating to the Arts of Painting, Gilding, Varnishing, Glass-Staining, Graining, Marbling, Sign-Writing, Gilding on Glass, and Coach Painting and Varnishing; Tests for the Detection of Adulterations in Oils, Colors, etc.; and a Statement of the Diseases to which Painters are peculiarly liable, with the Simplest and Best Remedies. Sixteenth Edition. Revised, with an Appendix. Containing Colors and Coloring—Theoretical and Practical. Comprising descriptions of a great variety of Additional Pigments, their Qualities and Uses, to which are added, Dryers, and Modes and Operations of Painting, etc. Together with Chevreul's Principles of Harmony and Contrast of Colors. 12mo., cloth. $1.50

PALLETT.—The Miller's, Millwright's, and Engineer's Guide.

By HENRY PALLETT. Illustrated. In one volume, 12mo. $3.00

PERCY.—The Manufacture of Russian Sheet-Iron.

By JOHN PERCY, M.D., F.R.S., Lecturer on Metallurgy at the Royal School of Mines, and to The Advanced Class of Artillery Officers at the Royal Artillery Institution, Woolwich; Author of "Metallurgy." With Illustrations. 8vo., paper. . . . . 50 cts.

PERKINS.—Gas and Ventilation.

Practical Treatise on Gas and Ventilation. With Special Relation to Illuminating, Heating, and Cooking by Gas. Including Scientific Helps to Engineer-students and others. With Illustrated Diagrams. By E. E. PERKINS. 12mo., cloth. $1.25

PERKINS and STOWE.—A New Guide to the Sheet-iron and Boiler Plate Roller:

Containing a Series of Tables showing the Weight of Slabs and Piles to produce Boiler Plates, and of the Weight of Piles and the Sizes of Bars to produce Sheet-iron; the Thickness of the Bar Gauge in decimals; the Weight per foot, and the Thickness on the Bar or Wire Gauge of the fractional parts of an inch; the Weight per sheet, and the Thickness on the Wire Gauge of Sheet-iron of various dimensions to weigh 112 lbs. per bundle; and the conversion of Short Weight into Long Weight, and Long Weight into Short. Estimated and collected by G. H. PERKINS and J. G. STOWE. $2.50

PHILLIPS and DARLINGTON.—Records of Mining and Metallurgy;

Or Facts and Memoranda for the use of the Mine Agent and Smelter. By J. ARTHUR PHILLIPS, Mining Engineer, Graduate of the Imperial School of Mines, France, etc., and JOHN DARLINGTON. Illustrated by numerous engravings. In one volume, 12mo. . . $2.00

PROTEAUX.—Practical Guide for the Manufacture of Paper and Boards.

By A. PROTEAUX, Civil Engineer, and Graduate of the School of Arts and Manufactures, and Director of Thiers' Paper Mill, Puy-de-Dôme. With additions, by L. S. LE NORMAND. Translated from the French, with Notes, by HORATIO PAINE, A. B., M. D. To which is added a Chapter on the Manufacture of Paper from Wood in the United States, by HENRY T. BROWN, of the "American Artisan." Illustrated by six plates, containing Drawings of Raw Materials, Machinery, Plans of Paper-Mills, etc., etc. 8vo. $7.50

REGNAULT.—Elements of Chemistry.

By M. V. REGNAULT. Translated from the French by T. FORREST BETTON, M. D., and edited, with Notes, by JAMES C. BOOTH, Melter and Refiner U. S. Mint, and WM. L. FABER, Metallurgist and Mining Engineer. Illustrated by nearly 700 wood engravings. Comprising nearly 1500 pages. In two volumes, 8vo., cloth. . . . $7.50

REID.—A Practical Treatise on the Manufacture of Portland Cement:

By HENRY REID, C. E. To which is added a Translation of M. A. Lipowitz's Work, describing a New Method adopted in Germany for Manufacturing that Cement, by W. F. REID. Illustrated by plates and wood engravings. 8vo. $6.00

RIFFAULT, VERGNAUD, and TOUSSAINT.—A Practical Treatise on the Manufacture of Varnishes.

By MM. RIFFAULT, VERGNAUD, and TOUSSAINT. Revised and Edited by M. F. MALEPEYRE and Dr. EMIL WINCKLER. Illustrated. In one volume, 8vo. (*In preparation.*)

RIFFAULT, VERGNAUD, and TOUSSAINT.—A Practical Treatise on the Manufacture of Colors for Painting:

Containing the best Formulæ and the Processes the Newest and in most General Use. By MM. RIFFAULT, VERGNAUD, and TOUSSAINT. Revised and Edited by M. F. MALEPEYRE and Dr. EMIL WINCKLER. Translated from the French by A. A. FESQUET, Chemist and Engineer. Illustrated by Engravings. In one volume, 650 pages, 8vo. (*Ready June* 1, 1874.)

ROBINSON.—Explosions of Steam Boilers:

How they are Caused, and how they may be Prevented. By J. R. ROBINSON, Steam Engineer. 12mo. $1.25

ROPER.—A Catechism of High Pressure or Non-Condensing Steam-Engines:

Including the Modelling, Constructing, Running, and Management of Steam Engines and Steam Boilers. With Illustrations. By STEPHEN ROPER, Engineer. Full bound tucks . . . $2.00

ROSELEUR.—Galvanoplastic Manipulations:

A Practical Guide for the Gold and Silver Electro-plater and the Galvanoplastic Operator. Translated from the French of ALFRED ROSELEUR, Chemist, Professor of the Galvanoplastic Art, Manufacturer of Chemicals, Gold and Silver Electro-plater. By A. A. FESQUET, Chemist and Engineer. Illustrated by over 127 Engravings on wood. 8vo., 495 pages. $6.00

☞ *This Treatise is the fullest and by far the best on this subject ever published in the United States.*

SCHINZ.—Researches on the Action of the Blast Furnace.

By CHARLES SCHINZ. Translated from the German with the special permission of the Author by WILLIAM H. MAW and MORITZ MULLER. With an Appendix written by the Author expressly for this edition. Illustrated by seven plates, containing 28 figures. In one volume, 12mo. $4.25

SHAW.—Civil Architecture:

Being a Complete Theoretical and Practical System of Building, containing the Fundamental Principles of the Art. By EDWARD SHAW, Architect. To which is added a Treatise on Gothic Architecture, etc. By THOMAS W. SILLOWAY and GEORGE M. HARDING, Architects. The whole illustrated by One Hundred and Two quarto plates finely engraved on copper. Eleventh Edition. 4to., cloth. . $10.00

SHUNK.—A Practical Treatise on Railway Curves and Location, for Young Engineers.

By WILLIAM F. SHUNK, Civil Engineer. 12mo. . . $2.00

SLOAN.—American Houses:

A variety of Original Designs for Rural Buildings. Illustrated by 26 colored Engravings, with Descriptive References. By SAMUEL SLOAN, Architect, author of the "Model Architect," etc., etc. 8vo. $2.50

SMEATON.—Builder's Pocket Companion:

Containing the Elements of Building, Surveying, and Architecture; with Practical Rules and Instructions connected with the subject. By A. C. SMEATON, Civil Engineer, etc. In one volume, 12mo. $1.50

SMITH.—A Manual of Political Economy.

By E. PESHINE SMITH. A new Edition, to which is added a full Index. 12mo., cloth. $1.25

SMITH.—Parks and Pleasure Grounds:

Or Practical Notes on Country Residences, Villas, Public Parks, and Gardens. By CHARLES H. J. SMITH, Landscape Gardener and Garden Architect, etc., etc. 12mo. $2.25

SMITH.—The Dyer's Instructor:

Comprising Practical Instructions in the Art of Dyeing Silk, Cotton, Wool, and Worsted, and Woollen Goods: containing nearly 800 Receipts. To which is added a Treatise on the Art of Padding; and the Printing of Silk Warps, Skeins, and Handkerchiefs, and the various Mordants and Colors for the different styles of such work. By DAVID SMITH, Pattern Dyer. 12mo., cloth. . . . $3.00

SMITH.—The Practical Dyer's Guide:

Comprising Practical Instructions in the Dyeing of Shot Cobourgs, Silk Striped Orleans, Colored Orleans from Black Warps, Ditto from White Warps, Colored Cobourgs from White Warps, Merinos, Yarns, Woollen Cloths, etc. Containing nearly 300 Receipts, to most of which a Dyed Pattern is annexed. Also, A Treatise on the Art of Padding. By DAVID SMITH. In one volume, 8vo. Price. . . $25.00

STEWART.—The American System.

Speeches on the Tariff Question, and on Internal Improvements, principally delivered in the House of Representatives of the United States. By ANDREW STEWART, late M. C. from Pennsylvania. With a Portrait, and a Biographical Sketch. In one volume, 8vo., 407 pages. $3.00

STOKES.—Cabinet-maker's and Upholsterer's Companion:

Comprising the Rudiments and Principles of Cabinet-making and Upholstery, with Familiar Instructions, illustrated by Examples for attaining a Proficiency in the Art of Drawing, as applicable to Cabinet-work; the Processes of Veneering, Inlaying, and Buhl-work; the Art of Dyeing and Staining Wood, Bone, Tortoise Shell, etc. Directions for Lackering, Japanning, and Varnishing; to make French Polish; to prepare the Best Glues, Cements, and Compositions, and a number of Receipts particularly useful for workmen generally. By J. STOKES. In one volume, 12mo. With Illustrations. . $1.25

Strength and other Properties of Metals:

Reports of Experiments on the Strength and other Properties of Metals for Cannon. With a Description of the Machines for testing Metals, and of the Classification of Cannon in service. By Officers of the Ordnance Department U. S. Army. By authority of the Secretary of War. Illustrated by 25 large steel plates. In one volume, 4to. . $10.00

SULLIVAN.—Protection to Native Industry.

By Sir EDWARD SULLIVAN, Baronet, author of "Ten Chapters on Social Reforms." In one volume, 8vo. $1.50

Tables Showing the Weight of Round, Square, and Flat Bar Iron, Steel, etc.,

By Measurement. Cloth. 63

TAYLOR.—Statistics of Coal:

Including Mineral Bituminous Substances employed in Arts and Manufactures; with their Geographical, Geological, and Commercial Distribution and Amount of Production and Consumption on the American Continent. With Incidental Statistics of the Iron Manufacture. By R. C. TAYLOR. Second edition, revised by S. S. HALDEMAN. Illustrated by five Maps and many wood engravings. 8vo., cloth. $10.00

TEMPLETON.—The Practical Examinator on Steam and the Steam-Engine:

With Instructive References relative thereto, arranged for the Use of Engineers, Students, and others. By WM. TEMPLETON, Engineer. 12mo. $1.25

THOMAS.—The Modern Practice of Photography.

By R. W. THOMAS, F. C. S. 8vo., cloth. 75

THOMSON.—Freight Charges Calculator.

By ANDREW THOMSON, Freight Agent. 24mo. $1.25

TURNING: Specimens of Fancy Turning Executed on the Hand or Foot Lathe:

With Geometric, Oval, and Eccentric Chucks, and Elliptical Cutting Frame. By an Amateur. Illustrated by 30 exquisite Photographs. 4to. $3.00

Turner's (The) Companion:

Containing Instructions in Concentric, Elliptic, and Eccentric Turning: also various Plates of Chucks, Tools, and Instruments; and Directions for using the Eccentric Cutter, Drill, Vertical Cutter, and Circular Rest; with Patterns and Instructions for working them. A new edition in one volume, 12mo. $1.50

URBIN.—BRULL.—A Practical Guide for Puddling Iron and Steel.

By ED. URBIN, Engineer of Arts and Manufactures. A Prize Essay read before the Association of Engineers, Graduate of the School of Mines, of Liege, Belgium, at the Meeting of 1865–6. To which is added A COMPARISON OF THE RESISTING PROPERTIES OF IRON AND STEEL. By A. BRULL. Translated from the French by A. A. FESQUET, Chemist and Engineer. In one volume, 8vo. $1.00

VAILE.—Galvanized Iron Cornice-Worker's Manual:

Containing Instructions in Laying out the Different Mitres, and Making Patterns for all kinds of Plain and Circular Work. Also, Tables of Weights, Areas and Circumferences of Circles, and other Matter calculated to Benefit the Trade. By CHARLES A. VAILE, Superintendent "Richmond Cornice Works," Richmond, Indiana. Illustrated by 21 Plates. In one volume, 4to. $5.00

VILLE.—The School of Chemical Manures:

Or, Elementary Principles in the Use of Fertilizing Agents. From the French of M. GEORGE VILLE, by A. A. FESQUET, Chemist and Engineer. With Illustrations. In one volume, 12 mo. . . . $1.25

VOGDES.—The Architect's and Builder's Pocket Companion and Price Book:

Consisting of a Short but Comprehensive Epitome of Decimals, Duodecimals, Geometry and Mensuration; with Tables of U. S. Measures, Sizes, Weights, Strengths, etc., of Iron, Wood, Stone, and various other Materials, Quantities of Materials in Given Sizes, and Dimensions of Wood, Brick, and Stone; and a full and complete Bill of Prices for Carpenter's Work; also, Rules for Computing and Valuing Brick and Brick Work, Stone Work, Painting, Plastering, etc. By FRANK W. VOGDES, Architect. Illustrated. Full bound in pocket-book form. $2.00

Bound in cloth. 1.50

WARN.—The Sheet-Metal Worker's Instructor:

For Zinc, Sheet-Iron, Copper, and Tin-Plate Workers, etc. Containing a selection of Geometrical Problems; also, Practical and Simple Rules for describing the various Patterns required in the different branches of the above Trades. By REUBEN H. WARN, Practical Tin-plate Worker. To which is added an Appendix, containing Instructions for Boiler Making, Mensuration of Surfaces and Solids, Rules for Calculating the Weights of different Figures of Iron and Steel, Tables of the Weights of Iron, Steel, etc. Illustrated by 32 Plates and 37 Wood Engravings. 8vo. $3.00

WARNER.—New Theorems, Tables, and Diagrams for the Computation of Earth-Work:

Designed for the use of Engineers in Preliminary and Final Estimates, of Students in Engineering, and of Contractors and other non-professional Computers. In Two Parts, with an Appendix. Part I.—A Practical Treatise; Part II.—A Theoretical Treatise; and the Appendix. Containing Notes to the Rules and Examples of Part I.; Explanations of the Construction of Scales, Tables, and Diagrams, and a Treatise upon Equivalent Square Bases and Equivalent Level Heights. The whole illustrated by numerous original Engravings, comprising Explanatory Cuts for Definitions and Problems, Stereometric Scales and Diagrams, and a Series of Lithographic Drawings from Models, showing all the Combinations of Solid Forms which occur in Railroad Excavations and Embankments. By JOHN WARNER, A. M., Mining and Mechanical Engineer. 8vo. $5.00

WATSON.—A Manual of the Hand-Lathe:

Comprising Concise Directions for working Metals of all kinds, Ivory, Bone and Precious Woods; Dyeing, Coloring, and French Polishing; Inlaying by Veneers, and various methods practised to produce Elaborate work with Dispatch, and at Small Expense. By EGBERT P. WATSON, late of "The Scientific American," Author of "The Modern Practice of American Machinists and Engineers." Illustrated by 78 Engravings. $1.50

WATSON.—The Modern Practice of American Machinists and Engineers:

Including the Construction, Application, and Use of Drills, Lathe Tools, Cutters for Boring Cylinders, and Hollow Work Generally, with the most Economical Speed for the same; the Results verified by Actual Practice at the Lathe, the Vice, and on the Floor. Together with Workshop Management, Economy of Manufacture, the Steam-Engine, Boilers, Gears, Belting, etc., etc. By EGBERT P. WATSON, late of the "Scientific American." Illustrated by 86 Engravings. In one volume, 12mo. $2.50

WATSON.—The Theory and Practice of the Art of Weaving by Hand and Power:

With Calculations and Tables for the use of those connected with the Trade. By JOHN WATSON, Manufacturer and Practical Machine Maker. Illustrated by large Drawings of the best Power Looms. 8vo. $10.00

WEATHERLY.—Treatise on the Art of Boiling Sugar, Crystallizing, Lozenge-making, Comfits, Gum Goods.

12mo. $2.00

WEDDING.—The Metallurgy of Iron;

Theoretically and Practically Considered. By Dr. HERMANN WEDDING, Professor of the Metallurgy of Iron at the Royal Mining Academy, Berlin. Translated by JULIUS DU MONT, Bethlehem, Pa. Illustrated by 207 Engravings on Wood, and three Plates. In one volume, 8vo. (*In press.*)

WILL.—Tables for Qualitative Chemical Analysis.

By Professor HEINRICH WILL, of Giessen, Germany. Seventh edition. Translated by CHARLES F. HIMES, Ph. D., Professor of Natural Science, Dickinson College, Carlisle, Pa. . . .

WILLIAMS.—On Heat and Steam:

Embracing New Views of Vaporization, Condensation, and Explosions. By CHARLES WYE WILLIAMS, A. I. C. E. Illustrated. 8vo. $3.50

WOHLER.—A Hand-Book of Mineral Analysis.

By F. WOHLER, Professor of Chemistry in the University of Göttingen. Edited by HENRY B. NASON, Professor of Chemistry in the Rensselaer Polytechnic Institute, Troy, New York. Illustrated. In one volume, 12mo. $3 00

WORSSAM.—On Mechanical Saws:

From the Transactions of the Society of Engineers, 1869. By S. W. WORSSAM, Jr. Illustrated by 18 large plates. 8vo. . . $5.00

www.ingramcontent.com/pod-product-compliance
Lightning Source LLC
LaVergne TN
LVHW010143110826
845151LV00002B/410

* 9 7 8 1 4 2 5 5 3 8 6 0 6 *